Plenty of Room at the Bottom

ASHOK BHOWMICK

"To see a World in a Grain of Sand
And a Heaven in a Wild Flower,
Hold Infinity in the palm of your hand
And Eternity in an hour."

—William Blake

ASHOK BHOWMICK

PLENTY
OF ROOM AT THE
BOTTOM

Epic Press

Belleville, Ontario, Canada

ISBN: 978-1-4600-0270-4
LSI Edition: 978-1-4600-0271-1
E-book ISBN: 978-1-4600-0272-8
(E-book available from the Kindle Store, KOBO and the iBooks Store)

Cataloguing data available from Library and Archives Canada

To order additional copies, visit:
www.essencebookstore.com

Epic Press is an imprint of *Essence Publishing.* For more information, contact:
20 Hanna Court, Belleville, Ontario, Canada K8P 5J2
Phone: 1-800-238-6376 • Fax: (613) 962-3055
Email: info@essence-publishing.com
Web site: www.essence-publishing.com

Printed in Canada
by

Acknowledgement

I am deeply grateful to the late Prof. Rooshikumar Pandya for motivating me to write this book and giving me continual emotional support while writing, at times listening patiently and suggesting, justifying my arguments. His incessant enthusiasm and encouragement has been a vital force for me in completing this work. May God let him know that the book is in print now!

Dedication

This book is dedicated to:

Friends in the corporate sector
Friends who want to do something new
Friends who like calculated adventures
Friends who can imagine the unforeseen
Children who have time to drive in a new world

Remembering my revered teachers—

Preface

The contribution of research in fundamental science creates an expansion of thoughts and ideas of engineering, which ultimately helps develop technology for the benefit of mankind. Technology is that all-pervading force that drives forward and creates new products that we can touch, feel and sense. Thus, a concept is transformed into something tangible by the dialectic between a scientific thought and fundamental laws of *nature*.

Kevin Kelly, founder editor of WIRED magazine and former editor of Whole Earth Catalog, has put forward such a new insight to look at technology as an underlying force. Perhaps, mankind apprehended technology as a self-dictating hidden power for the first time while learning to make fire by rubbing one stone with another. Technology is not to be confused with the high-end product like a gizmo or a spaceship or a transformer! Of course, technology is inherent in all of these as the invisible power that makes the product a complete self. However, technology is the total effect that is manifested through the functions. For example, if a screw represents engineering, the energy by hand that drives the screw is science then, the forward motion is technology.

The famous author Alvin Toffler narrated wonderfully long ago in *Future Shock* on what technology does for society. It accelerates our lives and in the process generates a shock. It is similar to the shock generated in the medium when you pierce a screw through it. Society gradually adapts to that shock and accepts the results. Let us call it the AAA factor—Acceleration-Adaptation-Acceptance. Just consider the evolution of the telephone. When Alexander Graham Bell invented it in 1876, it did not ring off the hook with calls from potential backers. President Rutherford Hayes said, "That's an amazing invention, but who would even want to use one of them?"

Coping with the acceleration and adapting to the shock takes less time when the benefits are visualized immediately and takes longer when the benefits are slow and somewhat distant. The process may also create conflict and stress because the common mindset seems to prefer the precedent to progress—the familiar to the new. This eternal looking backward instead of forward by millions is but the normal inertia. Just like when you pierce a screw through a piece of wood, the dead cells resist it—do not allow the screw to penetrate easily.

This mindset is reflected in our day-to-day conversation. Quite often we hear our elders say, "Oh, I hate this cell-phone." They might have heard their grandparents say, "Oh, I hate the telephone…"

But the inner joy of the journey is hidden therein—overcoming the resistance, protests and obstacles! It has always been so in the forward motion of our civilization throughout history. Indeed, it is a given that there is no turning back from this forward journey driven by technology. It is absolutely unidirectional and perhaps moves along the axis of time. You can't think of travelling in a horse cart in this century or communicating through telegraph or using a candle for your everyday lighting requirements.

The fact is, not everyone in society is directly involved in the practice of science, engineering or technology, but as we progress, everybody has to learn to adapt to the shock. Anything that assists you in adapting to the shock faster is of immense value. Information and knowledge help you do just that since generally speaking, an informed mind is better prepared and able to manage the change.

A book offering such information and knowledge could be targeted to:

1. Peer group or
2. People who are uninitiated or
3. People who have some sense of the issues but need some information/knowledge in enhancing their interests.

In each case, the style and substance of the book will vary.

My book *Plenty of Room at the Bottom* is meant for the third group but offers an update to the first group and insights to the seriously curious lay people of the second group as well.

It is also a book that offers a glimpse of where our civilization is rapidly heading…

The inspiration for this book came from Richard Feynman's famous lecture entitled *There is Plenty of Room at the Bottom* in the year 1959. In fact, the modern technological ability to practically manipulate atoms and molecules at nanometer scale, at which *nature* set its order, enables us to open up further new directions. Our ability to go by three orders of magnitude deeper than microtechnology is creating a gigantic amount of space to work on anything and everything. In this, we now have the capability to work directly with the fundamental, independent entity of the material world—the atoms!

The possibilities of new and unforeseen changes are beyond imagination. In the words of Nobel Laureate scientist (IBM) H. Rohrer who discovered the scanning tunneling microscope, "It might help to remember that micrometer had no significance for a farmer plowing his field with an ox and plow 150 years ago— nor for the ox or the plow. Nevertheless, the micrometer changed plowing—it is the precision standard for the tractor."

For the next 100 years or so, nanotechnology is going to evolve with a potential that might change our worldview and give us a new shock. In a sense, I see this book as a moral duty of helping the society to absorb and adapt to the shock by becoming informed in advance. Call this book a forecast, a precaution or even a warning!

I believe what is important is not a particular improved sunscreen lotion launched by a company using nanoparticles or a particular low wattage gizmo applying some nanosize transducer, but rather the conceptual understanding of the fact that with this new technology we could create things as *nature* does, that we could mimic the processes inherent in *nature*, that we can harness the energy crisis of the world with alternate clean sources of energy, that we can have ways and means to provide affordable healthcare for everybody on earth!

Ashok Bhowmick
Mississauga, Ontario, Canada
2014

Contents

CHAPTER I

Nanoscale and Nanotechnology

A person who has never made a mistake has never tried anything new.
—Albert Einstein

Time and again, science stalwarts confront the question towards the end of their formal career, '*What big might follow next in science?*' The question originates out of an expectation from the vision of a person having vast experience and insight. As of now, the constituents of matter down to the fundamental level of quarks and gluons are known; so are the laws governing the objects at galactic distances; all the fundamental forces are known; work for a grand unification is continuing for greater understanding of the origin of universe; man has known the principles and also advanced technologically to communicate and commute at the distances of planets; deciphered with great ability the knowhow of biological systems including the self; known the techniques of healing and safeguarding self from disorders and dysfunctions. Within these frontiers, what is the hope for something else or can there be a new direction to go ahead?

By itself, man's quest to look into the smaller is a major origin of important inventions. For example, it is Thompson's assertion of cathode ray as made of electrons and his conclusion in 1987 that all atoms contain some amount of electrons that has triggered the interest to look inside matter beyond the reach of naked eye. Subsequently, it followed through the work of Rutherford and Bohr to the structure of atom and finally the structure of the nucleus → nucleons → quarks, and gluons, all that are discovered through the involvement of great many ideas and technologies. Actually, this approach is instinctive to human nature, which in some sense like saying, "Break open to see what is lying there inside." History of science shows this impounding quest of mankind to discover *nature*, including the self, as the mother of all inventions and thus, is the expansion of our knowledge on the intricate architecture of *nature*. The science at nanometer scale is one most recent domain of such human endeavour.

The Greek word *nano* means *dwarf*. Obviously, Nanoscale and Nanotechnology indicate the sphere of activity that deals mainly with small objects or in small sizes. The word *dwarf* carries a sense of forceful reduction of size or a hindered growth. A similar sense is also in effect when we talk about artificial (man-made) and natural small objects. In fact, the introduction of the term classifying a separate area of activity follows from the terminology of dimension that scales by factors of 10.

Prof. Richard Feynman is acclaimed as the founding father of today's nanotechnology. He was the first person to give us a vision of benefit if we can go smaller, in his phenomenal lecture entitled *There Is Plenty of Room at the Bottom* at Caltech on December 29, during the Christmas of 1959. However, he never uttered the word *nano* in his entire discourse but rather encompassed all dimension as *small* in general. The non-imposition of specific limits in size defining the *small* essentially broadened the horizon of the domain from the smallest at the size of an atom to a somewhat larger self-sustaining device having typical size of a cell in living organisms. In his words, "*I am not inventing antigravity, which is possible someday only if laws are not what we think. I am telling you what could be done if the laws are what we think; we are not doing it simply because we haven't yet gotten around to it.*" Having uttered such a realisation, it was as if he gave an answer through an introspection of how much is really known of *nature* by how much mankind is actually able to create like *nature*. This also implies the apprehension of the fundamental knowledge necessary to understand the level of dimension at which nature really functions to make things.

Richard Feynman is perhaps the first man in the history of science who could distance himself to stand aside an entire kingdom of scientific knowledge and look back to say somewhat like, "Hey guys, look! We have too much of knowing *nature* and its rules; should we not try to create something new now as *nature* does, at least begin trying to inherit its architecture?" This metamorphosis in visualization is by itself revolutionary. The innate proposition of considering atoms as the fundamental ingredient in *nature's* toy-house (that we should be able to manipulate for our benefits) should actually be considered as a unique consciousness in the practice of science. Thereby, the play in *nature's* toy-house ultimately transmutes to creating and recreating the self; not by way of reproduction using *nature's* machinery, but making the fundamental machinery to make things as *nature* does! Even if such an epitome does not give birth to new physics, it surely points towards a new direction in modern science to build a whole new kingdom.

There are discussions on how much the modern development in nanotechnology is actually influenced by Feynman's premier lecture. Nevertheless, a one-to-one correlation in application is actually not important when we talk about a new idea. Significant is that even after so many years, particular scientific developments (through its own course of realization) can be traced back into the vision of a single mind spoken so wholeheartedly to people who practice science. At that time it might have appeared almost like a dream; nevertheless, that dream is now keeping the entire world awake!

Before proceeding further to describe the facts and figures of *nano*, it is pertinent to review how our experience of the physical world evolves as we go through the scale of dimension. In the following, the different levels of experiences are tabulated from our observations in Table 1A and 1B. Table 1A gives some examples of objects in nature as we go larger and larger, while 1B describes the sames as we go smaller and smaller in dimension. Don't worry if the chart appears difficult to read. Simply imagine it as a bridge that extends between the cosmic world and the world of atoms. Nonetheless, a scan along the variation of object experiences thereupon portrays the astounding similarity in the inner constructs of the two worlds.

This chart also illustrates the vast range of access of modern science, a span from 10^{-17} to 10^{23} (or 40 steps of 10) in meter scale, the typical metric length for our everyday use. Now, if we define a much finer scale to measure length as *nano-meter*, then our current yardstick would be equal to 10 x 01 x 10 x 10 x 10 x 10 x 10 x 10 x 10 times the new tiny stick of measurement. By doing so, we are actually shifting our origin in the length axis by 10^9 towards the left. In that, the same length domain now is re-enumerated to span from 10^{-8} to 10^{32}. The new yardstick is too small, though, as a reference in our daily life, but truly, *nature* does perform at this scale. Imagine an amoeba, the single-cell living being that causes serious health issues if it can somehow sneak into our stomach! Isn't it a practical length scale for such a small creature? The space within the stomach is infinitely large to this tiny creature. Human eyes cannot deal with such small an amount as a nanometer. So buying a 5×10^9 nanometer long item instead of 5 meters, for example, apparently does not make sense. However, if one is asked to think in terms of weaving a piece of cloth, the matter becomes different. The typical distance between the interwoven threads in a piece of cloth is about 100 micron, i.e. 10^5 nm (compare the chart). If you have thread of 1 nm diameter and the ability to manipulate it to weave at that scale, imagine how dense and tough that piece of cloth would be! Inventing a technology for such an action would give rise to a whole new chapter in the clothing industry! However, it still depends on the technology that is able to maneuver matter at nanometer length scale, in short, the *nano*-technology.

Table 1A
How large a size, seen from what distance

Scale by 10 Going large ↑	The nanometer scale	Example of object (or experience)
10^0 meter : 1 m	10^9 nano meter	Man is about 2 meter tall
10^1 meter : 10 m	10^{10} nm	Where the foliage of a leaf is seen
10^2 m : 100 m	10^{11} nm	Outer limits of a city garden
10^3 m : 1 km	10^{12} nm	Distance for parachute jump
10^4 m : 10 km	10^{13} nm	Typical small city limits
10^5 m : 100 km	10^{14} nm	Sight from satellite
10^6 m : 1000 km	10^{15} nm	Entire north hemisphere is seen
10^7 m : 10000 km	10^{16} nm	The size of Earth
10^8 m : 10^5 km	10^{17} nm	Moon's orbit
10^9 m : 10^6 km	10^{18} nm	Part of Earth's orbit
10^{10} m : 10^7 km	10^{19} nm	Orbits of Venus and Earth together
10^{11} m : 10^8 km	10^{20} nm	Orbit of Jupiter
10^{12} m : 10^9 km	10^{21} nm	The entire solar system
10^{13} m : 10^{10} km	10^{22} nm	Solar system looks small
10^{14} m : 10^{11} km	10^{23} nm	Sun appears like a star

Table 1A (continued)

10^{15} m : 10^{12} km	10^{24} nm	Sun blends off with the cosmos
10^{16} m : 10^{13} km : 1 Light Year (LY)	10^{25} nm	The dark of infinity
10^{17} m: 10 LY	10^{26} nm	Only stars and nebulae
10^{18} m : 100 LY	10^{27} nm	Inside the Milky Way
10^{19} m : 1,000 LY	10^{28} nm	Full extent of the Milky Way
10^{20} m : 10,000 LY	10^{29} nm	Periphery of Milky Way
10^{21} m : 100,000 LY	10^{30} nm	Whole Milky Way with other galaxies
10^{22} m: 1 Million LY	10^{31} nm	All galaxies look small
10^{23} m : 10 MLY	10^{32} nm	Reach of modern telescopes

Table 1B
How large a size, seen from what distance

Scale by 10 Going small ↓	The nanometer scale	Example of object (or experience)
10^0 meter : 1 m	10^9 nanometer	A leaf in the garden
10^{-1} meter : 10 cm	10^8 nm	Can delineate the structure within a leaf
10^{-2} m : 1 cm	10^7 nm	Internal structure in a leaf is visible
10^{-3} m : 1 mm	10^6 nm	Cellular structure within a leaf shows up
10^{-4} m : 100 micron (μm)	10^5 nm	Cells within a leaf can be defined
10^{-5} m : 10 μm	10^4 nm	Objects inside the cell are visible
10^{-6} m : 1 μm	10^3 nm	Nucleus in the cell is visible
10^{-7} m : 1000 Å	10^2 nm	Full extent of chromosomes within the cell
10^{-8} m : 100 Å	10 nm	DNA chain's size
10^{-9} m : 10 Å	1 nm	Several blocks within a chromosome together
10^{-10} m : 100 pico m :1 Å	10^{-1} nm	Electron cloud within the carbon atoms of the leaf
10^{-11} m : 10 pm	10^{-2} nm	Electron orbit surrounding the atomic nucleus

Table 1B (continued)

10^{-12} m : 1 pm	10^{-3} nm	The empty space between electrons and nucleus of an atom
10^{-13} m : 100 femto m	10^{-4} nm	The nucleus of an atom
10^{-14} m : 10 fm	10^{-5} nm	The nucleons in the nucleus of an atom
10^{-15} m: 1 fm	10^{-6} nm	The size of single proton
10^{-16} m : 100 Auto m (AM)	10^{-7} nm	The domain of quark particles that make the nucleons
0^{-17} m: 10 Auto m	10^{-8} nm	Science have not reached that small yet

In fact, technology is the knowhow that originates when understanding of fundamental science is transformed through the power of engineering manipulation into something that is directly beneficial to mankind and society. Crudely speaking, it is like saying, if you want butter, you've got to process the milk before! But first, you must have milk as a basic ingredient, and here, it is the fundamental understanding in the science underneath. Whether milk is to be boiled using an open hearth or in a sealed oven to make good quality butter is your task to engineer. It should be noted though that the quantity of butter always turns out to be less than the amount of milk used. Expenses, monitoring and dedicated work are essentials in order to process the milk. Likewise, nanotechnology is fundamentally dependent on our knowledge in deciphering *nature* at the nanoscale and our power to be able to manipulate *nature* at that scale.

The '*room*' in the title of Feynman's discourse surely meant the scope of doing scientific studies applying our existing knowledge provided we achieve a technical capability of dealing (not just seeing and realizing) at the bottom level of dimension down to the size of a single atom or molecule. Though, at times, parts of his discourse appeared like fairy tales, amazingly, fifty years down the line now, major portions of his propositions are appearing to be reality or near reality. It is not that everything in so called *nano* happened because he told it so; rather today's development appears as if it is a given and indeed been the natural course that modern science is supposed to go through. Importantly, he could foresee ahead of others, even though some inventors today admitted to have not known seriously about his discourse prior to their inventions. It would be relevant in this regard to examine the correlation of some of the cited examples in his discourse with the recent developments within the nano-bracket.

The modern microlithography and nanolithography can be correlated to his idea of writing with atoms—that is, documenting and printing things at a much finer and smaller scale so that the space available could hold more contents—an essential requirement for our advancing civilization that generates huge volumes of information every day for every next day. He imagined to ultimately scale writing down to atomic size. In fact, there are two approaches to this issue; one, from the top, i.e. improving the technology at hand to be efficient enough to go to the required finer level, ultimately to the stage of being able to write with atoms. This is categorized as 'top-down' approach. Otherwise, one can have a sophisticated machine that is capable of holding one atom after the other and place them on suitable writing pad and thus complete a sentence. This opposite approach is referred as 'bottom-up' approach and is somewhat like building a castle out of pebbles, but without any defect whatsoever. In fact, using a Scanning Tunneling Microscope (STM), scientists at IBM have demonstrated this possibility quite authentically.

The fascinating development of STM in the field of microscopy might be associated to Feynman's very strong persistence on improving the then-existing electron microscopes, proposing to make them at least a hundred times better in resolution. It is interesting to observe his mental stake on this machine; as if he already had the right kind of gadget to carry his vision into realization and didn't even need to imagine a new type of instrument except proposing further amenable sophistication onto it.

Conceptually a little different, the birth of STM should be marked as a revolution in the science and technology at nanometer scale. Nevertheless, the development of STM should be admitted to have been dependent on the parallel development of computer hardware and software, especially large memory, faster computing and image processing technique. Today, it is possible to image atoms and generate patterns with atoms on suitable substrates, though it is not that cost effective in practice. Having achieved the capability of manipulating atoms, Feynman urged physicists to involve in atomic scale chemical analysis that he visualized to be like working from first principle. In his words, "*It would be very easy to make an analysis of any complicated chemical substance; all one would have to do would be to look at it and see where the atoms are.*" Research studies on such direct approach is underway and has been partly successful as well. Having acquired the capability of maneuvering a process atom by atom, he proposed a chemical synthesis by a physical technique that is capable of simply putting the atoms where we want them to be for the new compound to be born! As he said, "*The principles of physics, as far as I can see, do not speak against the possibility of maneuvering things atom by atom. It is not an attempt to violate any laws; it is something, in principle, that can be done; but in practice, it has not been done because we are too big.*" Further, he anticipated the important issues as we try to get smaller by saying, "*The problem of resistance is serious*" and "*We have a lot of new things that would happen that represent completely new opportunities for design.*"

Since the world at this scale is governed by the laws of quantum mechanics, it is immensely important to know how exactly the different material properties essential for useful electronic devices modify as we go smaller in size. Also, what is the smallest possible actual limit in size that shows the property of bulk material? Both 'top-down' and 'bottom-up' approaches can be useful in deciphering this query. That is, we can chisel a bulk material down to the tiniest possible size and measure the changes in certain properties or we can go on adding one atom with another to sequentially grow in size in order to know the onset of any bulk property. For example, superconducting material may be used to circumvent the resistance issue in small circuits. However, one needs to know the minimum number of atoms required for a material to exhibit superconductivity. In other words, what is the smallest size of a cluster of niobium atoms that might exhibit superconductivity? Interestingly, research is being pursued using both the approaches in order to decipher the '*small*' at nanoscale, which means examining the nanoscale properties both by breaking the bulk into nanosize entities and as well synthesizing *clusters* of countable metal atoms in it. Therefore, Feynman's anticipation could be considered the genesis of today's molecular electronics, where specific molecules are designed as quantum electronic devices and synthesized using batch processes to assemble into useful circuits thereafter, using self-organization or self-alignment. Not just for molecules as active electronic components in a circuit, but he also said, "*We can use not just circuits, but some system involving the quantized energy levels, or the interactions of quantized spins, etc.*" Modern read heads for magnetic disk drives, which are actually an outcome of the discovery of Giant Magneto Resistance (GMR) spin-valve systems might be mentioned in this regard.

In 1959, the computing machines were so large that even schoolchildren today would find them as items for fairy tales or museums because we now have them small enough to be contained within the grip of our palms. Feynman indicated about such a situation explicitly and emphasized on the necessity of miniaturization of computing machines. Nevertheless, we are yet to achieve his scale of 'small' computing machines in which the wires would be 10 to 100 atoms in diameter, and the circuits a few angstroms across. Nanoelectronics is the domain where circuits are thought to be made using quantum dots and nanotubes. Apart from miniaturization of conventional electronics, 23 years later in 1983, he proposed a completely new conceptual domain of computing machine, the 'quantum computer,' in his lecture entitled *There's Plenty of Room at the Bottom, Revisited* at the Jet Propulsion Laboratory. He perhaps mentioned the difficulty that we are going to face when the computing machines are shrunk to atomic dimension due to the fact that nature at this scale is purely quantum mechanical. In his words, *"First, we can't use classical mechanics or classical ideas about wires and circuits. We have atoms, and we have to use quantum mechanics. Well, I love quantum mechanics. So, the question is, can you design a machine that computes and that works by quantum-mechanical laws of physics—directly on the atoms—instead of by classical laws?"* Quite obviously, such new devices cannot operate on conventional electronics, which is but conceptually classical.

However, before we appreciate the essence in his sentence, let us have a clear note of differentiation between classical and quantum mechanics. To put it across simply, mechanics is the relation between force and distance and it is established by the travel distance measured through the change in coordinate of initial position of an object because of the application of a known amount of force. The generalization of this measure leads to the formulation of the laws or principles of mechanics. The activities that you execute in your daily life since morning are all inherently based on the laws of classical mechanics. For example, riding on the lift, driving your car, eating your food etc. in which you have applied some amount of force and there is some amount of explicit or implicit displacement involved as well. Time is implicitly involved in the description of force and therefore, essentially mechanics turns out as the relation between displacement and time, i.e. x and t. In the classical domain of our everyday life, we can actually measure both time and displacement simultaneously and precisely. So the laws of mechanics in this domain are exactly describable and predictable. However, when your object becomes the size of an atom or you cross a certain dimension in length scale, it is not possible to measure both time and displacement simultaneously and accurately. The world becomes probabilistic, in which if you measure one parameter precisely, for the other you would only have a probabilistic measure within a certain error limit. This very fact changes the laws of mechanics, and this domain of mechanics with inherent indeterminism is called quantum mechanics, in which the different levels of probability (or options) are referred to as the quantum levels. Therefore, the generalization in this domain is not conventional because it does not directly correspond to our usual experiences of daily life and an understanding through correlation tolls on our imagination. *Nature* at the atomic level is undeniably best described quantum mechanically. It is indeed a matter of pride that people have understood this whimsy hidden deep inside *nature* and mastered the conceptual basis of it!

The problem in computing machines as we go smaller in size can also be realized through the observations laid down by Moore. Moore's law states that the size of computing machines shrink by ½ every 4 years. Extrapolation shows that by 2035, we should achieve the atomic scale. But in 2012, Moore's law should have been in trouble as the size was anticipated to reach the quantum domain. Performance could not be improved further simply by shrinking the size of the logic devices. Feynman proposed the new class of logic devices based on the principles of quantum mechanics, viz. reversible logic gates, if we really want to have miniaturized computing devices. In other words, the idea is to take advantage of quantum mechanics rather than be limited by it and develop a processor that simulates the natural phenomenon more naturally and exponentially faster than a digital computer. The concept may be simplified by putting it as follows: a 'digital bit' may only store information in the form of a sequence of "0s" and "1s," whereas a 'quantum bit' may be in a superposition state of "0" and "1," that is, representing both values simultaneously until a measurement is made. A sequence of N digital bits can represent a single number between 0 and 2^N-1, while N quantum bits can represent all 2^N numbers simultaneously. A quantum computer with only 300 quantum bits can represent a system with $2^{300} \sim 10^{100}$ elements, representing a number greater than the number of atoms in the observable universe today! A quantum computer should therefore be able to solve problems much more complex compared to that which a digital computer ever could.

Many physicists have been fascinated by the amazingly synchronized system in the biological world; the connectivity, parallel processing and the finest decentralized structure of management in complex biological organizations including human beings. But Feynman perhaps is the first person to think of converting his fascination by proposing to replicate the superb order in biological systems into the physical world, as man's creation. Here again, he knew the need to know biological systems more intimately and emphasized on improving upon the microscopes. Scanning probe microscope (SPM) has come up as the second generation STM and it is now capable of seeing directly and as well manipulating the biological entities on atomic scale. Feynman's vision however did not stop at just seeing atoms intimately and manipulating them. He proposed building tiny self-sustained complete machines like the molecular workshop in our body called *cells*. Data is acquired, processed, transported and even regenerated in the form of various signals all at a very tiny scale within the cells. For example, imagine yourself in a situation where suddenly an insect flies around you in an annoyingly attacking way; later peacefully think your actions stepwise, like in a slow motion movie, and be amazed by the processing of the entire episode in your brain in no time. Further, for example, the single cell amoeba is a living, active, and self-sustained creature, but as a whole, only a molecular system! Amazing! In his words, *"Consider the possibility that we too can make a thing very small which does what we want—that we can manufacture an object that maneuvers at that level!"* This is not an appreciation of astonishment but an introspective proposition of building things as is perceived from *nature*.

In this light, Feynman imagined making truly small machines that are also compact and complete and serve a function like in the tiny factories called *cells* in our body, which coordinate with a main system for final execution of any task. These tiny machines are proposed to be made by employing the kind of controllable master-slave robotic hands operating through a set of levers, similar in architecture to the master-slave arrangements used to handle nuclear materials in reactors; however, all are to be designed here on a very small scale. These tiny machines may be oriented to perform only a specific job or all might act the same way, like clones to perform different tasks. In analogy to large scale robots, by recent context these are referred as '*nanobots*.' In fact, Albert Hibbs suggested a very interesting possibility to him for relatively small machines. Though wild as an idea, these tiny machines can work as a mechanical surgeon if they are swallowed into the body to rectify a defective system, e.g. repairing a blood vessel or mending of a faulty heart. They dive in and 'look' around with a task to find out the fault and do the needful as might be commanded through external control; be it delivering a drug at particular place or slicing out an inner portion of the body. Other types of small machines might be permanently incorporated in the body to assist some inadequately functioning organ. Recent advances in medical science have actually scored a lot in this direction, specially laparoscopic and robotic surgeries, chip insertion technology for nerve disorders, besides improvements in smaller size pacemakers etc. In 1983, he reiterated this issue to utter, "*Suppose we could make free-swimming little gadgets like this. You might say, 'Oh, that's the size of cells—great. If you've got trouble with your liver, you just put new liver cells in.' But twenty years ago, I was talking about somewhat bigger machines. And he said, 'Well, swallow the surgeon.' The machine is a surgeon—it has tools and controls in it. It goes over to the place where you've got plaque in your blood vessel and it hacks away the plaque.*" As an idea it sounds absurd, but if accomplished, it would be a revolution. Once achieved, it will not only create impact for its promising applications in medical world, but also in material science. For example, in fabricating nanoscale devices by batch process where the tiny gadgets perform perfect precise jobs producing absolutely identical devices; analogically, such function resembles the marvelous task orientation observed in a colony of ants!

Eric Drexler has been the protagonist in the nineties fascinated by ideas of Feynman and did work to formalize the idea of making miniaturized devices with atoms. Perhaps he is the first person to coin the term 'nanotechnology.' In reply to a question, Feynman's son Carl said in 2005 that in his early days at MIT back in 1980s, Drexler was aware of his father's ideas and he did not think, except Drexler, any other scientist went into 'nanotech' by reading his father's ideas. Drexler extended the idea of fabrication by atoms as '*Molecular Engineering*' in 1981 with due reference to Feynman. Feynman's second discourse '*Infinitesimal Machinery*' in 1983, however, did not refer to Drexler's publication. *Molecular Engineering* largely focuses on computational designing of protein molecules. Though not lucid in reading, the intense speculations and language of presentation of this paper is far off to attract the attention of common readers into an entirely new domain of activity as compared to Feynman's discourse in the language of common people. However, Drexler's paper extends the visualization of *small-machines* through interesting analogical imaginations, e.g. *collagen* as cables, *sigma-bonds* as bearings etc. This paper thus is a very positive contribution towards understanding molecular systems from a different perspective, from the vision angle of '*infinitesimal machinery*' and perhaps should be considered as the first proposal to replicate the knowledge gathered in biological world into the physical world. In developing the understanding of quantum systems at nanoscale, Drexler's pioneer observations, e.g., "Like some enzymes, they can do work on reactant molecules to drive reactions not otherwise thermodynamically favored" are highly notable and important. Nevertheless, this paper also does not contain the word '*nano*' but it leads the readers to rediscover Feynman's '*Plenty of Room.*'

It may be worth to note that in 1980 the patent for STM was filed that received the Nobel Prize in 1986. But Drexler did not mention it, and neither did Feynman in 1983. The works on STM instrumentation and the studies done with it had largely been confined within rigorous scientific publications; it could be because it is difficult to judge the scientific merit of a new invention in its early years without getting actively involved in it. Drexler founded the *Foresight Institute* and published the book *Engines of Creation* in 1986 that has initiated interest in a large number of cross-disciplinary people and it has popularized ideas to shape as technology. These efforts could be marked as the premier liaison between *idea-to-laboratory* and *application-to-investment*—the usual path rendered to establish a technology. These moves from Drexler as the champion of Feynman's deliberation have certainly extended the domain of the thought process beyond the boundaries of people engaged only in science and hence it rediscovered Feynman. Richard Feynman expired in 1988, two years after STM received the Nobel acclaim, and Drexler is still continuing his contributions through '*Foresight*' and '*Nanorex*' with computer designs of APM and APPN that will be described in Chapter IV.

It is not important to judge how much the current developments in nanotechnology have been influenced by Feynman's lecture and Drexler's formalization because the same truth might appear completely independently from altogether different points of views, perspectives and investigations. However, it is opinionated that Drexler has created unrealistic expectations that have created untimely loads and threatened the field rather than aiding it. This might have introduced unwanted social hindrance towards the freedom of scientific research. But in today's world, full-fledged development of a technology cannot wait till it is finished at the laboratory; rather, a collective endeavor of the entire society is important for its realization. It may be that scientists engaged actively at laboratories normally sense directly ten to twenty years down the line, and Drexler's visionary approach is something that might shape up fifty to hundred years down the line, a period long enough in the future to risk of getting modified through unknown turmoil, upheavals, twists and turns.

1. The New Approach

It is not actually clear how the term *nano* replaced the term *small* or who initiated it, though three major inventions might be considered to be accountable in ushering in the remarkable sensation observed today. These are:

1. The discovery of the Scanning Tunneling Microscope (STM);
2. The development of the Atomic Force Microscope (AFM); and
3. The first ability to manipulate atoms individually in a controlled manner.

Gerd Binnig and Heinrich Rohrer discovered the STM at IBM's laboratory near Zurich. In fact they were looking for a better method to examine micro-defects in oxides. However, the concept was not accidental as has also been acknowledged by the Nobel committee in their report while honoring Binning and Rohrer in 1986. American scientist Russell Young at the National Bureau of Standards was first to envisage the investigation of a surface with the help of a suitable stylus. Conceptually it is similar to that of Braille reading where a blind person understands the letters by fingers. Here, the finger is substituted with a fine stylus and the vertical movement is recorded as data as the stylus scans across. The recorder information can be transformed into an image. The amount of details in the image vis-à-vis the resolution depends on the sharpness of the stylus and its ability to follow the structure of the surface. But an exceptionally sharp stylus having almost a single atom at the tip is prone to be damaged quickly, and moreover there is always a chance of accidental mechanical contact destroying the tip of the stylus. This problem was suggested to be circumvented by keeping the stylus at a constant small distance from the surface regulated by a servo mecha-nism controlled by the current. Young made such an instrument for the first time by using the phenomenon of field emission, i.e. applying a sufficiently high potential between the stylus and the surface, a current flows with a strength vary-ing with respect to the distance between the surface and the tip of the stylus. Instead of vertical distance here, the changes in the value of current flowing are recorded as data. In Young's instrument, the constant distance was kept at 20 nm. But the resolution was inferior to that of an electron microscope. Young recog-nized that much better resolution could be achieved by being able to utilize the quantum mechanical tunneling effect that requires a distance between surface and stylus of the order of 1 nm when a current is supposed to flow even at low volt-ages. However, Young could not overcome the experimental difficulties to succeed in converting his idea into practice. Binning and Rohrer solved these technical problems that finally gave birth to the Scanning Tunneling Microscope or STM. In this regard it is worth here to quote Binning's words in his Nobel lecture: *"We were often told it should not have worked in principle."* STM is not revolutionary in concept but created a revolution in its range of applications and in its ability to make mankind directly reach and sense the fundamental building blocks of mat-ter, the atoms.

It is said that Feynman was taken to see the STM at IBM's Watson Research Center in Yorktown Heights, New York. Scientists showed him the STM and said they could see atoms. Feynman corrected them and said they were observing tunneling of electrons. The Nobel document also says that STM is not a true microscope because it does not give a direct picture of the object. In order to appreciate the new attitude towards microscopy that STM has introduced, it is necessary to review the electron microscope before. Very justifiably, the 1986 Nobel prize in physics was shared with Ernst Ruska, who made the fundamental contribution towards the successful development of the electron microscope. His study of electron-optical properties of simple magnetic coils while he was a research student in Berlin has paved the way for imaging of objects using electrons. Ruska was under the supervision of Max Knoll, and he achieved a magnification of 15 times using two coils in series. Improving the details, he could make an instrument in 1933 that is regarded as the first modern electron microscope, having a resolution much higher than an optical microscope. Siemens appointed him, and the first commercial electron microscope was marketed in 1939. This event has been marked as a breakthrough. To date, the instrumentation in electron microscopy is continuing to achieve more with further higher precision. In this regard, it is also pertinent to differentiate between microscopy, electron microscopy, tunneling microscopy and force microscopy in order to appreciate the progress in science and technology at nanoscale that these successive developments have ushered in.

The limitation in directly observing objects of micro-dimension using light comes from the wavelength of visible light. Objects having dimension less than the typical wavelength of light cannot be resolved. This limits the resolution of microscopes using visible light to some 400 nm. Breakthrough came when it became possible to image an object using an electron beam. With the discovery that magnetic coils can function like an optical lens for an electron beam, it became evident that a divergent electron beam from a source can be focused to a point. According to quantum mechanics, any fundamental particle is associated with its characteristic wave, called a de Broglie wave. Therefore, a guided electron beam can also be used like the visible light, in order to produce the image of an object. Since the wavelength of electron is much smaller, it is possible to achieve a much higher resolution in direct imaging using a guided electron beam. Theoretically the resolution could be unlimited. However, according to Heisenberg's quantum mechanical principle, there is an uncertainty into the determination of position. This imposes a limit to the acceleration potentials, however superb the design might technically be. Nevertheless, high-end sophisticated electron microscopes nowadays achieve a resolution of 0.1 nm. Two types of designs in electron microscopes are available. The design that Ruska originally made is called a Transmission Electron Microscope or TEM. In this, the object to be imaged is taken as a thin section, and the electron beam goes right through it similar to what happens in an optical microscope. There have been several later developments, among which the Scanning Electron Microscope or SEM is the most prominent. Here, a very sharply focused electron beam strikes the object that emits secondary electrons, which are collected by an electron detector and the current is recorded. With the help of magnetic coils, the sharp electron beam is scanned all over the object and the recorded variation of secondary emission is used to build an image of the object. As the depth of focus is very large, three dimensional imaging is possible here as opposed to two dimensional section imaging in TEM. However, the resolution in SEM is much poorer compared to TEM. Hence, effectively these two instruments are complementary to each other to study a small object by electron microscopy.

1.1. THE STM

As outlined before, Scanning Tunneling Microscopy (STM) is a conceptually different method of imaging small objects. It is not true microscopy because direct imaging is not done in this technique. Actually, STM probes the surface density of states of a material using tunneling current. Now, density-of-state and tunneling are phenomena arising out of quantum mechanics, the mathematical language that describes *nature* in reality. It is however not absolutely necessary to know the physics underlying these phenomena in order to practice (or to do production oriented works) it for nanotechnology. But, an understanding of electron tunneling phenomena enables us to appreciate the merit in the discovery of STM. In fact, it is not the only instrument that has enabled the imaging of atoms for the first time. Individual atoms were imaged way back in 1950s using FEM and FIM. The novelty in the discovery of STM is in solving some of the technical hardships finally enabling an instrument that allows investigating virtually every metal surface under most atmospheres and at most temperatures while FEM and FIM were restricted to a few refractory metal tips such as tungsten, molybdenum, platinum or iridium under vacuum and at low temperatures only. Though conceptualized by Young in USA, the reason for the ultimate success at Zurich was the exceptional precision of the mechanical fabrications and the apprehension that vibration isolation is a prime necessity. Finally the microscope was built on a freely floating platform using heavy permanent magnets. Nowadays many improved designs are available with spring-loaded free suspensions. Vibration isolation is extremely important because the effective change in current due to quantum tunneling is very small and highly sensitive to environmental conditions. In fact, three technical issues are very imperative in obtaining good results out of a conceptually simple instrument like the STM.

1. An effective vibration isolation to avoid environmental interferences that might disturb the tip and the surface with respect to each other.
2. A mechanism that allows the coarse approach of the tip and the sample. The sub-Angstrom scanning movement of the tip over the surface that was for the first time solved by using piezoelectric ceramics.
3. Most important is the construction of the tip and how to make it. This is crucial to obtain a good resolution. In fact the resolution of STM is found to be much higher than was predicted by models at the time of its invention.

There are two modes in which an STM can be operated. Normally, it is operated in *constant-current* mode. The tunneling current is amplified here using a current amplifier to become a voltage which is compared with a given value. A negative feedback loop controls the position of the tip via the piezoelectric ceramic responsible for the vertical movement, approaching or withdrawing the tip from the surface in order to keep the tunneling current constant. An equilibrium distance is established between the tip and the sample and while scanning over the *xy*-plane a contour plot of equal current surface is obtained that can be displayed using a computer.

The alternative mode of operation is called the *constant-height* mode. In this, the tunneling current is measured while the tip scans on a constant plane above the sample surface. This mode significantly avoids one drawback of the *constant-current* mode. Due to the absence of any feedback, there is no finite response time that otherwise limit the scan speed. However, in this mode there is a danger of collision of the tip with any adsorbate or step on the sample surface that might severely damage the tip and affect the electronics as well. Further, due to exponential dependence of the tunneling current, the height information is not directly available here. Therefore, this mode is mostly restricted to scan atomically flat surfaces.

Electron tunneling is a functioning concept first described by Fowler and Nordheim in 1982. It happens when a conducting tip is brought very near to a metallic or semiconducting surface so that their electron wave functions overlap and a bias between the two can allow electrons to tunnel through the vacuum between them. Classically this cannot occur. The wavelike characteristic of electrons that made electron microscopy possible also gives rise to such an event. For low voltages, the tunneling current is a function of the local density of states at the

Fermi level of the sample. Variations in current as the probe passes over the surface are translated into an image. As discussed before, the tunneling current flows between the tip of the stylus that scans the sample and the body of the sample. Piezoelectric elements are used to control the horizontal movement of the stylus in two perpendicular directions so that the surface is scanned along parallel lines. This is why it is called a scanning microscope. Another piezoelectric element measures the vertical movement of the stylus. Very precise styluses are fabricated with single atom at the tip that consequently enhances the image quality to a great extent, giving a horizontal resolution < 0.02 nm and vertical resolution < 0.1 nm, thereby making it possible to depict individual atoms. This makes it possible to examine the structures on surfaces to atomic precision or equivalently, as if sensing the atoms at fingertips. This is surely a step forward towards Feynman's vision of being able to make things with atoms. Whether it has been invented without the knowledge of his words or following them, apparently it is the natural course of progress in science like *nature's* own invitation to search for the '*room at the bottom.*'

1.2. THE AFM

The Atomic Force Microscope (AFM) has been the next invention in the journey towards reaching the atom within our sensory perception and acquiring the ability to manipulate it for our chosen uses. Subsequently, a group of such microscopes came into existence as the derivatives of STM to cater to specific purposes among which the first one was invented by Binning himself along with Quate and Gerber. These machines are a further step forward in our approach to directly feel atoms at fingertips. The job of the fingertip is actually carried out by a cantilever that has very tiny and sharp end point almost close to the size of few atoms together if not a single one. This cantilever feels the atoms on the surface of a material literally through a very soft touch, and the information is processed by measuring the force that it senses while surfing along a direction. The microscale cantilever in AFM is typically made of silicon or silicon nitride and has an extremely sharp nanometer size tip that acts as the probe to scan a specimen. As the tip is brought into proximity to the sample surface, forces between the tip and the sample lead to a deflection of the cantilever according to Hook's law. Forces that are measured in AFM include mechanical contact force, van der Waals force, chemical bonding, electrostatic force, magnetic force, capillary force, Casimir force, salvation force, friction force etc. According to the force to be measured, different types of AFM are designed. For example, Magnetic Force Microscope (MFM) for the specimen in which the variation in magnetic force is measured.

Further developments have recently taken place and are continuing in which additional quantities can also be measured along with the force. For example, photothermal microscopy, scanning thermal microscopy, etc. are specific techniques based on the principles of AFM but aiming towards studying the changes under particular cause. Actually the technological developments of high class piezoelectric elements facilitating very precise movements have enabled this class of microscopes.

There are several different ways to measure the cantilever deflection as signals to covert into images. Most generally, an optical laser beam is focused on top of the cantilever (opposite to the tip end) and is reflected back into an array of photodiodes that produces the signal that is used to create the image. Whenever a deflection happens, the change in the signal is recorded. Apart from this, optical interferometry, capacitative and piezoresistive sensing are also used as preferable alternatives. The sample in general is mounted on a piezoelectric tube that moves along ± z-direction and the cantilever scans through x-y direction. In some designs 3 piezo-crystals are arranged in a tripod fashion where each one is responsible for the motion along a particular x, y and z direction. Such advanced design circumvents some issues in using a tube scanner.

In principle, the operation of AFM is no different from that of an STM. Like an STM, here too, in order to avoid any collision of the tip with the sample surface or any adsorbate on the sample surface, a feedback mechanism is designed that maintains a constant force and accordingly manipulates the tip-to-surface distance. Several designs of AFM have been developed at commercial companies and different in-house research labs. Essentially there are two primary modes of operation in AFM. The *static* (or *contact*) mode in which static tip deflection is used as a feedback signal thereby maintaining a constant force between the tip and the sample during scanning. Measurements in this mode are prone to have noise because of drift and its execution requires low-stiffness cantilever; this mode is mostly executed in contact where the overall force is repulsive and therefore is referred to as *contact-mode*.

The other mode of operation is referred as *dynamic mode*. Here, the cantilever is oscillated very close to its fundamental resonance-frequency (or a harmonic); the modifications in phase, amplitude, or frequency, due to the interaction between tip and the sample, are measured to provide the information. There may be several schemes in this mode of operation, e.g. frequency modulation, amplitude modulation etc. Variation in oscillation frequency provide information in frequency modulation scheme. Frequency can be measured with much higher precision and the use of a stiff cantilever makes it possible to approach very close to the sample without any snap-in; true atomic resolution can be obtained in this scheme if the measurement is performed in ultra-high vacuum condition. Changes in oscillation amplitude or phase provide the feedback signal for imaging in amplitude modulation scheme. This is good for measurements at ambient condition. Applied both *in-contact* and *non-contact* modes, it is mostly used to discriminate between different types of materials on surface.

There could however be a third mode of operation most suitable for biological or soft materials called the *trapping mode* or *intermittent contact* or *force mode*. A very stiff cantilever is oscillated close to the sample in a noncontact way where part of the oscillation traverses into the repulsive regime so that the surface is intermittently tapped. This improves the lateral resolution in soft samples. The problems in contact mode, such as drag etc., are virtually eliminated in this mode. An electronic servo adjusts the height to maintain a set amplitude of oscillation as the cantilever is scanned over the sample. A tapping AFM image is therefore produced by imaging the force of the oscillating contacts of the tip with the sample surface. This is an improvement over the conventional contact mode operation and the chance of surface damage is minimal here. Lipid bilayers supported on suitable substrates or single polymer molecules in liquid medium, e.g. 0.4 nm thick chains of synthetic polyelectrolytes, have been imaged using this mode of operation. Some of these results will be discussed at appropriate places later. Nevertheless, the discussion above highlights the exceptional capabilities of these new generation machines in literally reaching the bottom of the material world down to the fundamental limit of atoms, making us feel the atoms within our sensory perceptions.

The question now comes, how are we going to utilize the exceptional capabilities of these machines? Or to what extent will we be able to manipulate *nature* at the scale of atoms? Would it satisfy Feynman's vision of making things with atoms for the benefit of mankind? Or ultimately, would it generate any profit?

2. Applications and Tasks

In fact, if all the activities might be arranged within the *nano* folder, it should appear as a continuous journey so far as the progress of science and technology is concerned. Technology that has its base in the science of condensed matter has gone through a trend of achieving more smaller-faster-cheaper products, especially in the development of microprocessors. Computational solutions have become more and more numerically intensive and so also, the complexity in chemistry has increased manifold. Nevertheless, there are opportunities that go far beyond this narrow objective and in addition, there are possibilities for revolutionary invention as well. The new age machines described above would assist to probe locally down to the scale of atoms. No doubt, these are the essential tools of manipulation, and are a real boon to science for shaping tomorrow's technology!

The entire gamut of endeavour towards the new technology might be classified into two major groups. One, nanotechnology, i.e. development of a new technology using devices-objects-operations-performances that is all inherently in nanometer scale; two, nanoscale science and technology, i.e. interpolating the existing activities into the nano-domain or looking for the effects and applications if one parameter is reduced down to nano-scale sizes. Of course there will be boundary cases as well as overlap areas. A method or technique dependent broad classification may be done as follows:

1. Local probe methods, e.g. STM, SPM, TEM, SEM etc. by which individual and/or group of atoms can be manipulated.
2. Beam methods, i.e. modification, fabrication and assembly at nanoscale using electron or ion beam from external sources.
3. Computational methods, i.e. solving different physical and chemical issues at nanoscale, designing possible model systems on a virtual platform, indicating the problems that might appear during actual realization and explaining an observed phenomenon, numerically.
4. Nano-materials: Most of the nanoscale research activities today might be broadly encompassed into this category. All of material science activities in fact can become a part of research in the *nano* provided there is some beneficiary effect observed due to the reduction of any parameter (defining the material) down to nanosize or amalgamation of another nanosize material. Activities in textile, ceramic, paints, nanocomposites, plastics etc. all may be grouped in this category.

Most of the activities and achievements currently under *nano* actually are the outcome of research at nanoscale with a projection towards possible technological application. Transferring the outcome of these researches into technology is of course beneficial, but they are still way off from Feynman's vision of small systems in which devices and machines could be synthesized out of atoms. The actual achievement on that scale is still far reaching. However, all achievements to date are immensely important as foundation knowledge to build the final kingdom.

CHAPTER II

Fiction or Future—the Quotes

As our circle of knowledge expands, so does the circumference of
darkness surrounding it.
—Albert Einstein

H. Rohrer, the Nobel prize winner for the discovery of STM, said in a lecture, *"It might help to remember that micrometer had no significance for a farmer plowing his field with an ox and plow 150 years ago—nor for the ox or the plow. Nevertheless, the micrometer changed plowing—it is the precision standard for the tractor."* Likewise, today's endeavor using the super-precision and precious machines to get closer to atoms and record the image of their arrangement within materials should not be observed as only expensive research. Where and how science pays back is always beyond the limits of human predictions. Forty years back, a radio used to mostly look like a rectangular wooden box fitted with two large knobs over a nice piece of linen cloth. The set usually would occupy quite a substantial portion of the drawing room, connected to some kind of antenna placed outside the house or inside the room. Something closer to what we are still used to seeing for our TV sets today. Turning those radio knobs would tune to the stations at particular frequencies and while passing from one to the other, these old valve sets would produce funny sounds. As a boy, I used to be fascinated and imagine the sounds as signals from extraterrestrials; I loved to listen to those sounds more than the transmission from any particular channel. But look at radio today! Within your fist and maybe as small as your shirt button! The smallest TVs have come down to the size of a wrist watch! For next generation kids, it may be funny to learn that a dish used to be required as an antenna to watch television!

Let us look into some of the results obtained through the use of special machineries developed to discover the world of *nano*. As discussed earlier, due to the dominance of quantum mechanical effects, each nano-entity having specific size and shape is distinct even when they are made of atoms of the same material. Therefore, they can be distinguished as independent individuals and identified by some name and surname, for example, *size*, or (typically) how many atoms are there, and *structure*, i.e. how the atoms are arranged with respect to each other. For example, carbon nanotube (CNT), whose very name suggests that it has nanometer size and the carbon atoms are so arranged that it creates a tubular structure (intricacies apart). Here, for example, the nanoscale entity C_{60} bucky-ball is understood to be made of 60 carbon atoms but arranged like a soccer ball. Though both are made of carbon, they are very different entities in terms of their qualities. Extreme caution is therefore important while probing them with sophisticated tools in order not to perturb their properties. Local probes like STM, AFM and SPM are like *fingertips* to interact with these individuals similar to the way we sense macroscopic materials with our fingers. Locations and properties as well as functions and associated processes, all may be sensed and conditioned through local probes. Interactions between the probe and the object depend on the active size of the probe and the distance in between. Such an approach is capable of accessing the scale of homogeneity down to atomic precision, which is unique and most attractive. The state of the art is that there are two controls corroborating with two different types of interactions; one that controls the distance between the probe and the object and the other that looks through or measures or does the study or acts as the tiniest eye as we convert the data into the form of images. Nevertheless, a very strong interaction between the probe and object might modify the characteristics of the object itself that sometimes limits the choice of an object and the probe. Some outstanding observations are cited below that certainly qualify the extreme capabilities of this class of machines.

In 1990, Donald Eigler of IBM Almaden and E.K. Schweizer from Fritz-Herber first moved 35 Xenon atoms to spell out the three letters of the IBM logo atop a crystal of nickel (picture available freely on IBM website). The tininess in this writing is incredible! Normal sized letters here in this book are about a millimeter, i.e. a million nanometers across. That means, writing at nanometer scale may be able to stuff close to one million letters into a millimeter. One can then imagine the amount of information that can be stored within a media of one square inch in size that has a capability of holding such writing!

The IBM website also has an example of a decorative pattern made with iron atoms atop a copper (111) crystallographic surface. Not only the image but also the structure is constructed using an STM. It is a corral structure created by plucking and placing individual iron atoms with an STM tip. Such engineering at atomic scale is a highly sophisticated work of art down to the finest achievable scale. The manipulation of atoms to form such a barricade structure has the purpose of understanding the fundamental physics in a tiny human generated pattern. Copper, being a highly conducting metal, has groups of electrons at its surface that are free to move. In scientific terminology in general, they are referred as 2D electron-gas. The electrons should be partially reflected when an iron atom is successfully placed on top of the surface. As a result, such a confined construct will produce undulations at particular locations on the remaining part of copper surface, more intensely within the enclosure. This is a phenomenon purely quantum mechanical in origin and this type of constructs are termed as 'Quantum Corral structure.' It is referred as *Stadium Corral* because of its shape. Don Eigler at IBM has been the key person in creating such wonderful nanoscale creations that gave scientists the access to decipher the reality of finite objects at a new length scale. Fig. 1 demonstrates the formative stages of a corral structure created by putting atoms one after the other to form a circular barrier. From the left hand corner image, following horizontally across the figure, it reveals how randomly spread atoms are arranged sequentially to form an ordered pattern that creates regular wavy undulations like that which happens on still surface of water as a stone is pelted vertically onto it. In fact, the pattern of undulations does vary with the geometric shape of the barriers as shown in Fig. 2. Barriers having corners create more interesting concentration zones possibly due to overlaps. Both hexagonal and square shows regular complicated patterns.

But, iron atoms could also be put in a different way on the same metal (copper) surface without creating the quantum corral structure. The Japanese Kenji character meaning *atom* could be written by iron atoms on a copper (111) surface. Such constructs containing only a few countable numbers of atoms are highly significant from the point of view of writing in atomic scale. The full extent of this written character is only a few nanometers across (size of an iron atom ~ 0.13 nm). A detailed description of the interactions of wave functions at the interface is a hardcore physics issue and surpasses the purpose of this book. Many such incredible developments are being reported on a day-to-day basis at IBM depicting the extreme capabilities of STM in engineering nature at the deepest level.[*] Atoms or molecules other than metal are also being used to fabricate interesting nanostructures. For example, Fig. 3 shows a man-like drawing done using STM

---

[*] Interested readers may visit the IBM website by Googling "IBM STM".

entirely with 28 carbon monoxide (CO) molecules on platinum surface. This CO-man is a construct that shows the limits of our permissible whimsy in modern science because it is only 5 nm or $5*10^{-9}$ m in length!

It is mandatory however to mention here that these experimental works are extremely sophisticated and require exceedingly low temperature and perturbation-free environment. Atoms and molecules are not obedient folks and never at rest! Even at absolute zero (-273°C) they vibrate within a limit (zero point energy). These experiments therefore require low temperature conditions obtained by using liquid helium (4 Kelvin ~ -277°C) to arrest atoms' translational and rotational degrees of freedom and substantially reduce the vibration so that they are locatable using an STM tip within a small spatial boundary. External influence caused by any sort of disturbance cannot be tolerated. Even a loud clap might pollute the measurement through the transmission of sound waves. These factors make the performance of such measurements extremely sensitive, eventually sophisticated and lengthy in process. Huge amount of time and patience are required in pre-conditioning the experiments in terms of bringing down the vacuum level (~ 10^{-11} to 10^{-12} millibar pressure) and attaining a steady thermal condition. Also, sustaining a stable condition without any external interference throughout the time scale of performance is a commendable task that involves very careful engineering in designing the experimental setup. Therefore, use of certain modern technologies is inevitable and needs to be exercised with zero tolerance limits. Such hardships turn the entire business, both time and money wise, expensive. Relieving one or the other parameter, for example, thinking of fabricating a nanostructure at room temperature, is still a challenge because most atoms or molecules will just not stay put! The research laboratory at IBM Zurich has recently succeeded in positioning a certain type of molecule at room temperature. The challenge here has been to find a suitable molecule that is slippery enough to be pushed around by an STM tip but sticky enough to remain in place after the withdrawal of the tip. The chemical bonds of the molecule should be strong enough not to be broken or altered as the molecule is pushed around. Fig. 4 shows a nanostructure done at room temperature by putting 11 rows of C_{60} molecules on a copper surface where each row contains 10 molecules, thus forming a filled rectangular arrangement. It could be possible because the C_{60} bucky-ball molecule is by itself stable at room temperature. Using the STM tip as a stick, any C_{60} molecule could be dragged left or right along the rows within the structure creating somewhat like an abacus pattern. Individual molecules make beads having less than one nanometer diameter forming the tiniest abacus in the world!

As discussed in the previous chapter, AFM is one step further advanced than STM that can also image at non-conducting environment. Huge amount of works are being carried out around the globe using AFM. Just to enunciate its capability, some prominent achievements could be mentioned here (with due courtesy to respective authors). For example, single strands of the polymer poly_2_vinylpyridine molecules are imaged at different pH using an AFM by operating the instrument in tapping mode under water (as medium). Single chains are found to be a minimum of 0.4 nm thick and the molecule conformation changes drastically in case the pH of the solution is varied even by a small amount. The implications of such intricate investigation is appreciated when we visualize the fact that polymers are all pervading in our modern society in various forms, like polyester, polythene, plastics and fabrics, etc. to an extent that our days could hardly be realized without the use of any polymer. Certainly we make plenty of things using polymer but it is mostly used in some kind of *condensed-matter* form. Properties of many naturally obtained polymers are well known even to the extent of the molecular arrangement and structures. However, we still lack in the knowledge of how to control them at single strand level or, more precisely, at molecular level. For example, what are the basic factors that govern the interweaving in polymer strands that makes a fabric? How it becomes tougher or fragile, denser or lighter? Is it possible to design at a fundamental level by engineering upon individual strands (or at molecular level) than what *nature* provides us? This work, showing the critical behavior of the conformation of a polymer depending on the conditions of the medium it is actually in (which here is either acidic or basic), is an extremely important step forward in building the understanding. This kind of study provides input at a fundamental level towards achieving the sought after technological goals.

Nevertheless, man's inquisitiveness as mother of invention takes pace once the right tool is in hand. This indulgence is perhaps not a different instinct than what prehistoric people might have done as they invented the first basic tool required for survival and growth! The motive behind it is somewhat like, "Explore! What the tool is capable of doing; if good enough to kill a wild pig...can it kill a tiger too?" Our activities in the civilized world are more composed, correlated to specific sets of knowledge that already exist and are recorded systematically for future references.

Four years down the line in 2009 it has been possible to outstandingly resolve the atomic anatomy of a single molecule using AFM. Yes indeed! It is something beyond imagination to see molecules directly, which have thus far been only a mental construct made out of correlation between observed facts and theoretical understanding. August 2009: researchers at IBM Zurich have made a major breakthrough. For the first time the anatomy of a complicated organic molecule is imaged with spectacular resolution, showing the arrangement of atoms with respect to each other; a structure that chemists had so far been speculating and calculating in order to create its plausible construct (i.e. what it should look like). Fig. 5 shows the image. This fascinating result that literally sets our eyes on top of the molecule stands out as a remarkable evidence of correctness of science in understanding the molecular structures! The imaging required an atomically sharp metal tip that contains only one metal atom which is (deliberately) terminated with a CO molecule to obtain the optimum contrast in imaging. Such an intelligent technique enabled the researchers to measure at short-range regime of forces, where sample-to-tip distance could be made as small as 0.5 nm, allowing the inner structure of the molecule to be visible. Even the positions of hydrogen atoms in the 1.4 nm long pentacene molecule are visible. The measurement has been performed at ultra-cold temperature -268°C. The success of the technique of this measurement also substantiates the fact that the foremost molecule at the tip actually governs the AFM contrast and resolution.

The significance of the achievement is that it initiated the potential to actually track Feynman's vision of *Direct Chemistry* into a near-reality. Feynman envisioned in his first lecture: "*The theory of chemical processes is based on theoretical physics. But chemistry also has analysis. If you have a strange substance and you want to know what it is, you go through a long and complicated process of chemical analysis. But if the physicists wanted to, they could also dig under the chemists in the problem of chemical analysis. It would be very easy to make an analysis of any complicated chemical substance; all one would have to do would be to look at it and see where the atoms are….*" His issue on electron microscope has now been overcome by atomic force microscope. With such achievement we are practically one step ahead in deciphering an unknown substance directly by dissecting the anatomy down to atomic precision; a capability that is opening up a new direction in doing chemistry!

Nevertheless, Feynman's outlook on atomic scale chemical synthesis that he phrased as, "*But it is interesting that it would be, in principle, possible (I think) for a physicist to synthesize any chemical substance that the chemist writes down. Give the orders and the physicist synthesizes it. How? Put the atoms down where the chemist says, and so you make the substance. The problems of chemistry and biology can be greatly helped if our ability to see what we are doing, and to do things at an atomic level, is ultimately developed—a development which I think cannot be avoided,*" has to still wait to be realized even though the advances are on right track. In this connection it is pertinent here to recall another very important work published in March 1999. The work reported on the measurement of the force required to rapture single covalent bond using AFM. It requires a force of about 2 nano-Newton to rapture the Silicon-Carbon bond and 1.4 nano-Newton to break the Sulfur-Gold anchor on surface. These kind of studies together with absolute control on the technique, keeps the promise to pave the way for new chemistry that Feynman envisioned.

The important question is what do we learn through all these achievements or what does the tough endeavors and remarkable results teach us towards the fulfillment of the actual goal of making things at nanometer scale? Do these skills bring in a new dimension in photography where STM and AFM are to be counted as the new age cameras capable of taking pictures down to atomic scale? The answer might be hidden within the fact when one perceives the inner-order in *nature* inherently decorated within the results. Having revealed such truths, one can comprehend that it is not only an element of pleasure in knowing the *nature* but also the fortitude of the scientific community to decipher the intricacies of *nature* to inherit it for beneficial purpose.

Slowly, we are indeed achieving the capability to engineer at the scale of atoms, the ultimate building block of matter so that we could follow through the process that *nature* executes. Nobel laureate Horst Stormer rightly puts it as, "*Nanotechnology has given us the tools…to play with the ultimate toy box of nature——atoms and molecules. Everything is made from it…The possibilities to create new things appear limitless.*" It is universally true that new things are not created through grace from the sky but germinate on the foundation laid through the understanding of old things. In science, in fact, both the activities go parallel; the volume of one type of activities might exceed the other at times. Contemporary efforts and activities as shown in earlier examples, are in fact inputs in building the foundation, e.g. setting the correct pillars by digging into our store of existing knowledge in order to make the Stormer's mentioned 'possibilities' into reality. Ronald Hoffmann, another Nobel laureate, rightly said, "*Nanotechnology is the way of ingeniously controlling the building of small and large structures, with intricate properties; it is the way of the future, a way of precise, controlled building, with incidentally, environmental benignness built in by design.*" As such, it might be observed that all natural systems and materials have foundation at nanoscale. An ability to control matter at a molecular level actually means the competence of tailoring the fundamental properties, processes and phenomena. This gives the power to create materials with desired properties having specific functions. The question comes, how are all these research activities on materials at smaller and smaller sizes going to be technology? Is it not too speculative to be far reaching? Maybe not, but an oriented endeavor might consume more time. As a matter of fact, it is worth it to introspect on 'technology' itself or review through facts as to what and whom it actually stands for! In fact, in the society we are so deeply within technology that it is difficult to feel. Somewhat like being contained within the universe, we fail to apprehend its dimension. Or being within atmosphere, we cannot normally sense breathlessness until withdrawn from it!

Therefore, what is technology? Is it a thing? Is it a certain class of activities or some specific domain within our belonging? Having a nice mobile phone or a large LCD TV or the most luxury car, etc.? Kevin Kelly asked this fundamental question intriguingly and perhaps illuminated the best on it. He propounded a new insight of introspection. In reality, technology is something that human beings have used since their primordial lifestyle. For example, stone gadgets in the form of weapons to hunt for food! That should be recognized as the birth of a specific idea driven by a particular necessity. This creation of special things useful to serve a purpose for better survival or to improve the quality of life is the prime motive in technology. What it has ultimately given the human being is a power, i.e. a force to strike with, leading to a confidence that later became inevitable so that life had come to a stage not manageable without it. Therefore, it is not the gizmo, nor the act. It is that invisible entity always with the individual and with the society as well, what *technology* is. Ancient people then learned how to make fire, a major invention in the life of the human species that culminated into the capacity of making our own food separate from what animals consume. Thereafter, we learned to clear the jungle, plow the land, and sow the seeds, marking the beginning of agriculture. Work → experience → knowledge → transformation in society for better survival leading to further domestication; there is an invisible force flowing through and passing onto the next generation. This force has empowered the human being and driven him to be segregated from animals. This force is constantly designing the intricacies within society and re-fabricating its structure at every moment of our existence. This is *Technology*. In a TED talk, Kevin Kelly explained it the best. To summarize the concrete facts that he has put up: *the very term Technology was first formalized through a US presidential address in the year 1952 and before that there had hardly been any use of the word within common people. The first use of the word dates back to 1829 where a course in arts and crafts was called as technology. Since the primordial period till date the continuing process has so empowered human being in every aspect of their life that today if withdrawn, people won't survive may be even for days together; simply because our body cannot consume raw food like animals and our cells cannot do photosynthesis like plants either! Over generations of survival in our formal structure, we have become so used to the entire process that we hardly care for it on a daily basis, till we come across something new and also unusual.* What has been the implication of this tremendous force that technology is? Its unleashed power has essentially transformed the place that we live and made us who we are. As Kelly rightly points out, this power has led us to invent ourselves in that, making us the most preciously domesticated species on earth. Take examples of all the things that each one of us comes in contact with, the inanimate objects, throughout the day. It is not difficult to find that all are basically through our inventions. The landmark makings are: the invention of wheels, recognizing metals, discovery of electricity and magnetism, understanding what materials are fundamentally made of, recognizing the electron and its use in making purposeful devices, i.e. electronics, understanding the power of gravity to explore outer space, and knowing the fundamental particles that the cosmos is built of. Most of our everyday actions around the globe, one way or the other, might be identified as the empowerment achieved through our experiences in these fundamental sectors. Whatever we have made for us—our home, the building, the water flow system, toilet, shaving blade, shoe, tie, watch, butter, toast, car, road, office etc.—technology is omnipresent. Unfortunately, we use the phrase only when we see a fancy gadget, uncommon at that moment. The truth is, we are always attached to the invisible threads emanating out of this force. As a spinoff, possibly, we invented a system of exchange called "money."

It is interesting to note that this force has also enabled us to go smaller in size at every step of progress, as far as the use of material is concerned, as we walked out of the stone-age; for example, shaped stone pieces as weapons → sized precise metal pieces as spears and axes → gunpowder, i.e. grains of materials. Notice as the size of material has reduced, it has become more effective in power! The strength we achieved at every step has enabled us to chisel further at the next step, like a parallel discovery, e.g. throwing a large stone with the force of hand might injure a wild buffalo, but it is not necessary because even a small stone can be thrown at sufficiently large speed to have the same effect. Therefore, a gadget became the necessity that could multiply the hand force several times to throw a small stone at great speed; it is easy now to define the gadget as a gun and the stone as a bullet! The inner understanding that came through for the learned people is that the momentum (mass * velocity) is most important, neither the size nor the mass of the stone! At the same time, this experience also generates a sense that there has been large wastage of materials in the previous step of technology, e.g. so many effective small pieces could be made from one large piece of stone! So many bullet balls could be made out of 1 kg metal! Similarly, at every step of our progress we have become more precise in hitting the target using gradually smaller pieces. For example, the knife, a common one has been the usual in ancient societies to cut tree branches and chop meat as well; now, we have innumerable varieties of knives, all precise and meant to serve specific purposes. As a matter of fact, neither butter knife nor Swiss army knife is capable of killing an animal! The urge to be more specific and precise, together with the sense of overuse of materials, is a genesis to go smaller. This actually imposes a universal direction underlying the *progress* in society because of the momentum imparted by technology. Therefore, technology, unlike science and engineering, is a vector quantity. In analogy, it might be visualized like a screw motion, where the screw itself is science, the rotation is engineering, the motion is technology, and the screw advances, i.e. science goes deeper, technology evolves like the motion. Certain attributes that this motion has given us are: (1) Diversity, e.g. think of all different knives! There are innumerable ones; sole objective is cutting but look at the difference between the butcher's one and the brain surgeon's one; (2) Options, e.g. your opportunity to select as per your purpose. Remember, in medieval times, doctors involved in surgery used to be referred also as butchers; imagine the contribution of special knives as surgical tools in giving a separate identity to the profession! (3) Acceleration of all aspects human life (maybe even for the domesticated animals!) The true account of this third attribute is given excellently in the book Future Shock by Alvin Toffler. His portrayal is still vivid, 43 years now since its publication in 1971, and as well true for all coming time in future. He visualized the universal journey of mankind, the bulk material referred as society. Though readers are recommended to his book, it might be pertinent here to quote some relevant portions just to reaffirm the sense in acceleration. For example, *"In the coming decades, advances in all these fields (technology proposed developments in society) will fire off like a series of rockets carrying us out of the past, plunging us deeper into the new society. Nor will this new society quickly settle into a steady state. It, too, will quiver and crack and roar as it suffers jolt after jolt of high energy change. For the individual who wishes to live in his time, to be a part of the future, the super-industrial revolution offers no surcease from change. It offers no return to the familiar past. It offers only the highly combustible mixture of transience and novelty"* (p. 217).

But at the same time there is also the observation, *"The greatest and most danger-ous marvel of all is the complacent past-orientation of the race, its unwillingness to confront the reality of acceleration."* It is still persistent! Scientific community encounters this issue almost on a daily basis. Think about the mindless, fact-dis-torting opposition against genetically modified food in view of an obvious crisis of food in near future. Also about anything new; so also is there against nano-technology! It is mostly because the race becomes afraid and hence rigid to a fur-ther adaptation with another pace of new acceleration, though it is inevitable. Again, Toffler writes, *"Thus man moves swiftly into an unexplored universe, into a totally new stage of eco-technological development, firmly convinced that 'human nature is eternal' or that 'stability will return.' He stumbles into the most violent rev-olution in human history muttering gist, that 'the process of modernization have been more or less completed.' He simply refuses to imagine the future"* (p. 215.)

It cannot be any different with nanotechnology! The prefix *nano* is only a cer-tification to man's ability to go another order of magnitude smaller than microtechnology (which is in use all the time now), to the extent of playing (or shooting) with the fundamental building blocks of matter, the atoms. It might as well be said that in order to save the world from many kinds of material and resource overuses and making the earth as a whole more habitable further into the destiny, nanotechnology proposes the indomitable new route. Though there will be jolts and rapid acceleration demanding serious adaptation in lifestyle and even thought processes, imposing control on freedom, benefits will be much more to circumvent all the doubts of today!

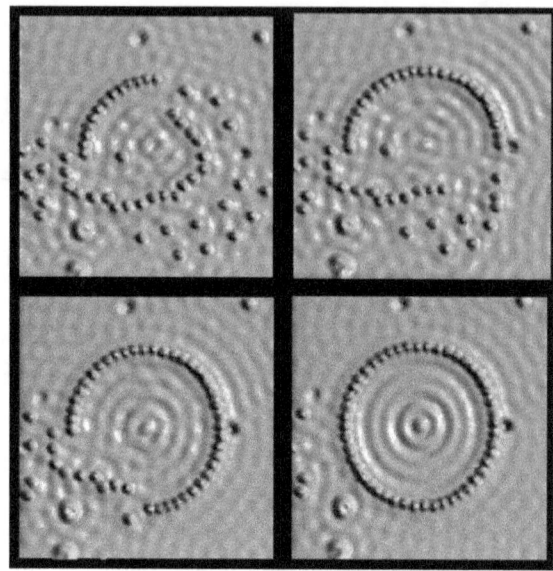

Fig. 1. Manipulation of iron (Fe) atoms on (111) copper (Cu) surface using the Scanning Tunneling Microscope (STM) TIP; the images show the stages in succession in the construction of a circular quantum corral structure. (Crommie, Lutz and Eigler, Images originally created by IBM Corporation; http://researcher.ibm.com/researcher/)

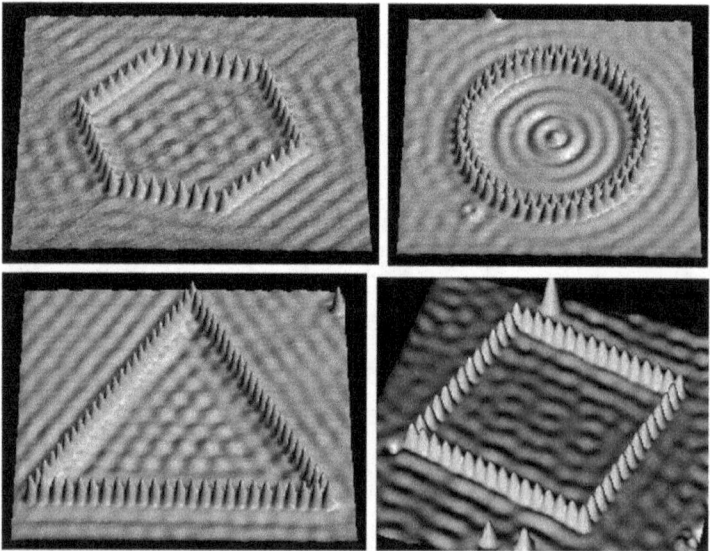

Fig. 2. Images of the creation of different shape quantum corral structures using Fe atoms on Cu surface. More complex patterns are generated within specific geometric shapes. (Crommie, Lutz and Eigler, Images originally created by IBM Corporation) http://researcher.ibm.com/researcher/

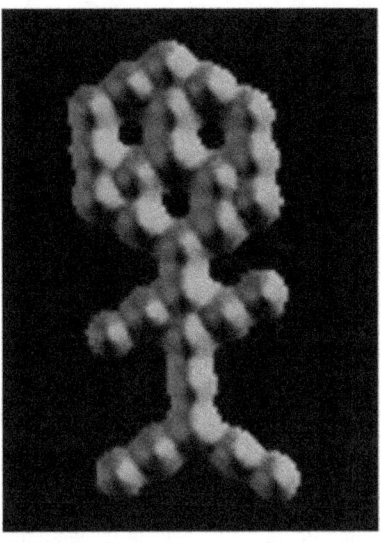

Fig. 3. The construction of a CO-Man on Platinum (111) surface using STM;
(Zeppenfeld and Eigler, Images originally created by IBM Corporation)
http://researcher.ibm.com/researcher/

Fig. 4. The Molecular Abacus: Constructed by manipulating C_{60} molecules on Copper (Cu)
surface using STM; (Courtesy of IBM Research – Zurich)
http://www.research.ibm.com/topics/

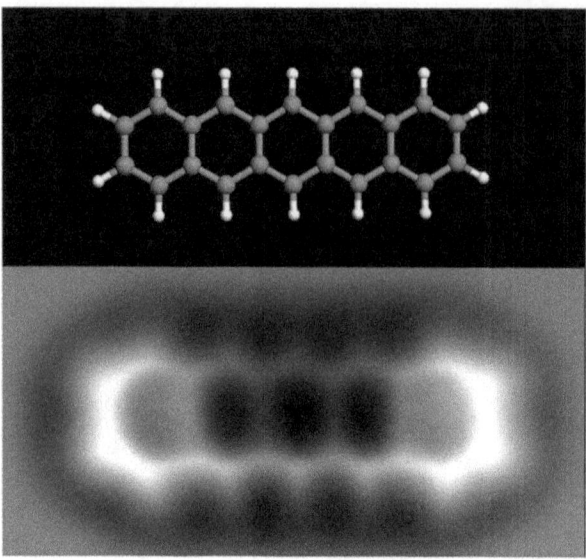

Fig. 5. The anatomy of single Pentacene molecule revealed through noncontact Atomic Force Microscopy (AFM); The image on top shows the ball and stick model of the molecule which under the AFM, is seen as shown in the bottom image (Courtesy of IBM Research— Zurich) http://www.zurich.ibm.com/imagegallery/

CHAPTER III

Dream, Vision, Investment, Promise & Stake

I think that only daring speculation can
lead us further and not accumulation of facts.
—Albert Einstein

1. A Change in Perception

We in modern society make plans, every year, every month, every week, every day and sometimes on hourly basis. Plan, plan and plan everywhere, every quarter and at every step. Why do we at all make plans? It is in order to achieve a task within a stipulated period. That means, we mostly know *what* to do and organize on *how* to do by making plans. Common folks normally, do not ask *why* a thing should be done, rather presume that the work has some relation somewhere with the main course of activities. Mostly, the motive of the work is supposed to be with the authorities. For all intents and purpose, this question remains unnecessary in our modern life till one hasn't been compelled to know the answer to serve a specific requirement. This fact is very common in our everyday routine. We measure a success by the sum total of work done in terms of financial profit and not by satisfaction of achieving a dream triggered by a query of *why*.

There is an age old dictum of *golden sphere* that might be enunciated in this regard. In society, most of the people know what they are doing, fewer know how they are doing it, and in fact a few know why they are doing certain thing. Three groups of people can be formed in this context which is actually contained concentrically. Easiest is to imagine three concentric spheres, one within the other, (1) forming three layers together. The lower group people who just perform a job they are asked to do, reside at the outer core, called the *bronze sphere*. People who in addition have the expertise and know how to perform the job belong to the middle sphere referred as the *silver sphere*. The smallest group of people, those who know *why* something is being done, are the core people contained within the inner-most *golden sphere*. The conventional (2) logical flow of work in an organization or company is, *what* → *how* → *why*, an inward direction. The *what-people* at the outer circle only do the task in hand, the job, and work for money. They do not get involved to extract anything on the far reaching consequences of the job done and are not really motivated by any root cause because they do not identify themselves with the system. People within the second inner circle have satisfaction over performance, i.e. in applying their knowhow to the extent possible to monitor the completion of a work that is being carried out by the *what-people*. But they also are not necessarily required to create work or research or find out an alternate or a more efficient mode of *how* to execute a job. Even though the people in this group are identifiable, they are not entirely integrated into the system. Ultimately, the inner group people belonging to the core are actually the system people. Either the system belongs to them or they belong to the system. Members of this group either direct or create the functions meant for the lower groups. In (3) such structural dissemination, the core people have to be always vigilant and keep track of the outputs from lower levels that gives rise to a whole set of managerial exercise in the process, which are accomplished by employing additional liaison manpower placed at the interfaces who work as *information-manipulators* to transmit information to the core group. Success in this type of conceptual model could never be satisfactory because most of the people in the system are not wholeheartedly contained in it. Their happiness may be accounted only with financial gain and not with their inner pleasure of accomplishment that an artist enjoys in performing his art. The manipulators, the so-called managers in the system, positioned at interfaces, always demand more incentives because the core group turns out to be functionally dependent on them. Facts are registered as what the managers convey to them and not as what they could have observed otherwise. Systematically, over a period of time, the motive disjoints from the *control*, sometimes to such an extent that *control* seldom carries any bearing on the motive. On the contrary, if carefully observed, it might be noticed that practically the approach is reverse in the functioning of any great leader or scientists or personalities or a very successful company or organization around the world. It has indeed been a self-managerial approach where the logical flow of work has been the opposite, i.e., *why* → *how* → *what*, an outwardly direction! (4) If a great world leader or organisation is taken as an example, it might be observed that their functioning is conceptually different, driven by an inherent cause at the nucleus. (5) In fact, personalities who lead have dedication to the cause, confidence to achieve and revered honesty. They deliver their belief in the beginning; in the process their dedication is expressed, their honesty is communicated and their confidence is rewarded. What they transpire are not instructions, rather the belief and a transcription of their own dream, which is accepted by the likemind or even by an empty-mind to be followed over a period of time; not just to do work, but to orient them towards the belief. When asked or hired, these like-minds change mindsets and don't work for money but toil for a cause. It may not be that everybody comes to the fore, but those who are on-board need no vigilance, need no manager to monitor. They manage themselves towards the objective with reason. (6) Therefore, it is said, "employ somebody just for a job, you get the work for money; but hire someone who believes in what you believe, your dream is shared and your employee self-integrates to toil for his own reasons and sweat till the end of his strength." No extra managerial vigilance is necessary because everyone becomes inherently a self-entrepreneur of his own faculty. They don't run after the so called *success*. On the contrary, they win, and *success* runs after them.

2. A Look Back at Technology

Together with the appreciation of the necessity of a changed attitude in order to welcome a new perception, it could be pertinent also to analyze the actual contribution of technology to the society. Back in 1958, a famous essay of American Founder and President of The Foundation for Economic Education, Leonard E. Read taught us that in today's civilization, nobody is capable of making, for example, even a pencil independently! In fact, there is nothing to be surprised with this truth. Person who puts the graphite rod into the wooden blockade does not make graphite. Production of graphite involves a lot more knowledge, money and manpower. Thereafter comes the making of particularly shaped graphite pieces. Therefore, even though it is apparently one person who made a factory to make pencils, the final product is an outcome of the collective wisdom and effort of mankind culminated through the process of selective exchanges over a considerably long period of time. That's where we stand in today's civilization! As the British journalist Matt Ridley (author of several books on popular science) brilliantly put forward in a TED Talk; all of us who work in society, in fact, work for each other and the amazing word *Exchange* preserves it. This *Exchange* is indeed like the spider on the web that builds the virtual net and we very precisely know, *business is synonymous to contact and communication*, that sustains entirely onto this net. To be precise, we are elements of an invisible *net*; interlinked for survival, at all moments, in every walk of our life! To reiterate Matt Ridley, this exists since our primordial days of hunting, as human beings learned to divide the labor in order to share the food, e.g. women used to gather wood logs for fire to get the share of meat brought by men! This is perhaps the learning when we started realizing the hidden teachings in the reproduction process, i.e. to create, one needs to share and, if the seed lies in one, the soil is in the other! In fact, this learning paved the way of differentiation between human being and animals. Quite agreeably, we have taken up *Exchange* as necessity even before we have taken up agriculture! It is therefore quite rational to say that human beings probably learned trade before agriculture; not due to the purpose of utilizing the surplus, but for self-survival! Hence, *Exchange* is a fundamental asset in entrepreneurship; be it in hunting and selling the meat or making clothes or the nanotubes! Matt Ridley emboldened this issue through a base example of a piece of stone tool compared to a computer mouse. Undeniably, both are technology! However, the former is obtained by single individual's endeavor whereas the latter is a result of huge collective action, to the extent that it is really not possible for any single individual to make it or frame an independent entrepreneurship to make it out of the scratch (i.e. from ground zero). That's where we are as a civilization, compelled to compromise more and more to work collectively as we progress in technology. Each one of us in effect works for each another, *share* and *exchange* in order to advance further. The primary contribution of technology inevitably is that it has established a more intricate relationship of interdependence recognized in the form of collective wisdom. In simple words, nobody can survive alone! There lies the hidden glue in the word *ADVANCE*: Association, Development, Verification, Amalgamation, Necessity, Compromise, Exchange! It is like saying, associate to develop and verify, then amalgamate with necessary compromise built by exchange.

3. Reinventing Management

The discussion in the first paragraph suggests a nonconventional management approach involving multidisciplinary scientific personalities to expand large scale nanotechnology programs. The second paragraph proposes a new function modality for entrepreneurs involved in such activities in which profiteering has a more evolved paradigm of estimation. Further, management of business and management of research have two different approaches. At the developing phase, business through an approach of research management might seem more plausible.

The purpose of this chapter is to emphasize that it is hardly sufficient to establish and sustain a new empire only through the foresight and speculations of great personalities who are quoted earlier. The empire has to survive ultimately through people in terms of useful products and potential business. Ten years down the line from the published IWGN report, many investments are done and the stocks are generated with immense expectations. At the same time, new anxieties are nurtured on probable consequences from the unknown and unchartered hazards thereby giving rise to ethical concerns. Quite legitimate! The history of societal evolution teaches us that when society encounters a new paradigm, invariably there arises the fear of the unknown which is ultimately overcome in time through experience and accepted through adaptation.

4. Redefining Investment

It is not essential to review government, venture capital, public and private investments in nanotechnology prior to 2005 as by now they are mostly done away with. Nevertheless all are available free of cost on the internet for anyone interested in having a look. It might yet be suitable to point out that the overall scenario has too many colors sprinkled in it to mislead anybody to assume that there is the golden opportunity of making quick bucks as a return! The overall scenario did not project if there could be any trend whatsoever to continue over a long period.

Things seem to have taken a different turn after 2005 from the natural course of seeking information and gaining knowledge. The business landscape changed to an extent possibly because of some amount of careless and out of proportion propaganda on the possible deliverables of nanotechnology; this in turn inculcated an over-expectation and a strong smell of money. Reports, information, analysis on investment, everything readily converted into sellable items under glossy covers, in the pretext of directly bringing business into the new domain of technology. In the name of instating discipline, such information packaging is rather to be seen as the first hindrance on free growth of a new discipline, which is but a precondition to germinate knowledge! A new depraved attitude has slowly crept in that cast a sense of monopoly on business on the far reaching products by imposing restrictions on vital information necessary for a free global growth of knowledge through inventions. Unfortunately, such an attitude has become so dominant that even a section of science practitioners apparently seems to have become infected! The sanctity of free growth of knowledge is contaminated by out of proportion expectations. Nevertheless, it should be kept in mind that wisdom in today's world has a collective nature, and realize at the same time that the new subject is still tied by an umbilical cord with the fundamentals of science!

Apart from all the growth curves on investments available on the internet, it is something else that requires a review—a review on the definitions of investment and outcome or deliverables and profiteering. What does a *stock* in nanotechnology mean? This should be the second most important definition after the definition of *nanotechnology* because in reality all applications eventually have to be accountable by taxpayers terminology. As such, it is hard to define. Even for science practitioners, it is highly nontrivial to predict what eventually will stand out as a valuable product. It is not possible to forecast or even develop an insight simply by scanning research publications and documents. As a matter of fact, a major area of science (except the sub-atomic domain) might be mapped onto the nano-domain in one way or the other! To elucidate, let me for example cite two recent publications of one of the champions in nano-world, Prof. Paul Alivisatos, UC Berkeley, who has also been a prime personality in the US government's national nanotechnology initiative commenced by former President Bill Clinton:

1. *"Transition from Isolated to Collective Modes in Plasmonic Oligomers"*— by Hentschel, M.; Saliba, M.; Harald, G.; Alivisatos, A.P.; Liu N. published in *Nano Letters*, 2010, 10 (7), pp. 2721-2726.

2. *"From Artificial Atoms to Nanocrystal Molecules: Preparation and Properties of More Complex Nanostructures"*—by Choi, C.L. and Alivisatos, A.P. published in *Annu. Rev. Phys. Chem.*, 2010, 61 (7), pp. 369-389.

These are very important, fundamental studies, apparently having implications in future technology.

However, it might be relevant to observe, it is indeed difficult even for a physicist or a chemist of a different specialty to interpret these articles right away, in terms of any future benefit towards nanotechnology. A document in this format might not be appropriate to recommend right away to any entrepreneur for a commercial exploitation. There might be hundreds of such examples created every day around the globe in the studies of fundamental science which builds the foundation for technological applications. An analogy might be given, for example, in terms of the diagram introduced by Feynman to represent the interaction between fundamental particles. To a particle physicist, a Feynman diagram communicates the hidden information. But if such a thing is put up as a painting in an art exhibition, will it really attract an art connoisseur to buy it? There is, therefore, a requirement of decoding the information into a suitable liaison-format for an investor or manufacturer to understand and judge upon the feasibility within the potential of his entrepreneurial endeavor. Apart from this, there is also the domain of computational nanotechnology, especially the kind that Eric Drexler had pioneered. The apparent first comment for realizing those things should be, *Far reaching!* Nobody knows how and where to materialize them! In summary it might be relevant to observe:

* Fundamentals in each topic has a level of maturity before engineering could be operational; there is no other way but to be patient on that account.
* It is necessary to create a liaison manpower who should have the mind-set of business but training in scientific research.

Incidentally, a special type of individual entrepreneurship like a *job-on-contract* course could be initiated for interested individuals having work credentials in an organization or a company. To some extent, this approach is like the job of freelance patent writers who have science background but are trained in law and writing. The initiative should though be flexible enough not to be attached to either the manufacturing company or the research organization. It might eventually evolve as a viable new career option essential to interface for a systematic transfer of knowledge to develop useful end products out of research results. Such career domain has to be interdisciplinary that might, for example, be commenced through an overlap of the post-docs in fundamental science to business school management graduates. For a new entrepreneurial drive aiming towards nanotechnology products, such manpower could be more in demand than only the management graduates. Building and sustaining an oriented growth thereby demands more thoughtful discussions across the table among stalwarts from multiple disciplines. Maybe, all of these have been told in various forms before by many people. Nevertheless, I would like to remember the French author and Nobel Laureate, André Gide in this regard who said, "Everything has been said before, but since nobody listens, we have to keep going back and begin all over again."

5. The Current Categories

The vast span and huge diversity of the research works under *nano* makes it non-trivial to impose any particular categorization. Nevertheless, a broad grouping can be followed in terms of the objective and target of deliverables.

* *Nanoenablers*: people who are involved in research, development, and manufacturing of sophisticated instruments, equipments and tools, targeted or designed for nanoscale investigations, are essentially the contributors to this category.

* *Nanomaterial Researchers*: this group contains people and activities that produce the substances different from the usually known ones (or sometimes the interpolated ones, if not a completely new one) that are useful for one or more of their properties. The special properties might be identified as a result of reduced dimension in nanometers. This block is quite large and it is difficult to isolate this group by any sharp boundary. However, this group is precious due to prospects of quick business.

* *Nanotechnocrats*: the original messengers who are the flag-bearers of Feynman's idea of constructing nanometer scale devices atom by atom. This group is by and large theoretical or computational (virtual) experimentalist. Modelling the design of nanoscale devices or functionality is the prime activity here.

Quite obviously, there cannot be any sharp line of demarkation among these categories. They are supposed to be interdependent and will thereby overlap in terms of academic interests and possible deliverables.

5.1. NANOENABLERS

These are the companies who started their activities very early in close connection and collaboration with researchers at academic institutions following the developments at laboratories. It should be noticed that each of them has one or the other special design or manufacturing capability. This is actually the group who has taken commendable risk and imparted tremendous courage to the scientific community; thus welcoming the new paradigm of thoughts in the process. It has been risky as all initiatives required heavy investments; there has neither been any promise of definite return of profit in the long-run, nor a possibility of immediate gain. Even if there is any, it is for the next generation! The work has to deal with too many unknowns that can only be corrected and consolidated through multiple tireless failure experiences. However, the gain is, once achieved it creates a brand for at least next 20 to 50 years. But, progress demands adaptive upgrades to maintain the standard and consistency; financial return is realized only through many years of a patient waiting and learning. The scenario is akin to that of the Gold Rush where second generation settlers benefitted and the first generation diggers fought against arrows. On the other side, the drive of the first generation made the shovel makers of that period rich!

A few prominent names in nanoenablers category are:

1. JEOL: An established company in manufacturing microscopes for many years. They have developed automated scanning probe microscopes (an improved variant of AFM, discussed in Chapter II). (www.jeol.com)

2. VARIAN: The company has now become a part of *Agilent Technologies*. They are leader in magnetic imaging and a good number of ultrahigh vacuum equipment. These are haute technology instruments that could be applied into nanotechnology research. However, the instruments are versatile enough to be applicable in researches other than nanoscience. (www.varianinc.com)

3. ARRYX: It is now a wholly owned subsidiary of Haemonetics. They have patented an optical trapping device that is also referred as an optical tweezer. This instrument utilizes the principle of momentum transfer associated with bending of light through an object. These optical traps can hold, move, rotate, join, separate, stretch, and otherwise manipulate hundreds of microscopic and nanoscopic objects, ranging from the size of a human cell down to less than 1/1000th the diameter of a human hair. Thereby, it has exclusive scope in exploring the *nanoworld*. (www.arryx.com)

4. Nanolnk Inc.: There are a whole lot of patents filed by this company on Dip Pen Nanolithography (DPN) technique and also license agreement with several universities. DPN is a cutting edge technology in top-down nanofabrication and is an already established method of fabrication at nanometer scale. Materials are deposited onto a surface via a sharp probe tip and the molecules are transferred from the tip onto the surface. The tip could be a pyramidal scanning probe microscope tip, a hollow tip or even a tip on thermally actuated cantilevers. Controlled deposition of molecules over surfaces can be performed even through a water meniscus. In fact, controlled transfer of molecule as *ink* was initially developed at Northwestern University in Professor Mirkin's group. In analogy to the macro-technique of a quill-pen, they were the first to introduce the term Dip-Pen Nanolithography. It is now proved to be one of the most promising top-down techniques of nanofabrication. (www.nanoink.net)

5. NanoNex: A company that has become synonymous to machines capable of producing finest achievable nanoscale imprints. Prof. Stephan Chou, from Princeton University, established this company

in 1999 and HITACHI was a primary investor in his drive. Today it has achieved the capability of providing user-friendly nanoimprint lithography (NIL) tools and solutions to a broad spectrum of markets for both experts and non-experts involved in micro and nanofabrication. As quoted in their website, the products include NIL tools, resists, masks, and also, processes for a variety of NIL, e.g. thermal and photo-curable NIL, and direct imprinting of materials. (www.nanonex.com)

6. ZYVEX: They introduce themselves by saying, "Zyvex commercializes nanotechnology to address real-world applications." They claim to focus on short and long-term nanotechnology applications with high-growth potential. The company has a very intense track-record of progress over a very short span since its commencing in 1997 by James R. von Ehr II and has now become synonymous to the term *nanomanipulators*. Their remarkable product line consists of 4, 6 and 8 position probing systems targeted for applications in semiconductor industry especially for quantum dot or nano-dot fabrication and these are able to manipulate samples at sub 100 nm dimension range. This company has wide range of alliances and collaborations with academic, business and government projects among which the NIST-ATP program is prominent. They have patented almost each of their new designs and development. It perhaps appears to some extent excess than required for any single line of development. As a matter of fact, there could be quite a few potential design alternatives of these instruments. However the design of a nanomaniputator is prioritized by the particular aim in an investigation. Therefore many times the universal designs require to be modified to cater to narrow focus. Nevertheless, having a core design always saves time and money. On the other hand, too many patents on successive improvements actually overshadows the academic willingness and flexibility. (www.zyvex.com)

There are many other companies enabling research and development in nanotechnology around the world, especially in Germany and Japan. For example,

7. Omicron Nanotechnology GmbH: They build custom design sophisticated complete instruments for in-situ measurements, like UHV STM/AFM with special nanomanipulators or semi-robotic hands to manipulate the process in-situ. (www.omicron.de)

8. Hitachi, Japan: An establishment for quality electron microscopes for several years now. They have recently launched the variable pressure scanning electron microscope that has lots of promise for nanotech research. (www.hitachi.com)

These are just a few examples of some investments that have been successful and each one could establish as a brand of facility providers for nanotechnology research. Nevertheless, history of science shows that the concept of new tool, equipment and instruments are born and initially takes shape at laboratories for the necessity of performing a new research. Things might not look polished, totally sufficient and completely optimized at the first success of results, but through international scientific publications it gets discussed, criticized and suggested for improvements or followed through improvements by somebody somewhere in the world, to cater to a different or a better purpose. The knowhow remains academically free and might be utilized without prior permission or payments. That's the merit of an academic scientific document over a legal scientific document called patents. Knowhow transferred to the company folder or expertise shifted out of the academic enclosure attains polished attire in the name of products; things become shining and compact. But specific knowledge gets entrapped within a legal shell as package for the market. The driving force changes as a kind of contest to survive which in the world of exchange (business), is measured as the *stock*. However, once we start measuring success in terms of a pile of specially printed papers that carries an exchange value, many other issues crop up, e.g. moral, ethical, environmental etc. standing as obstacles and diverting the attention from the real challenges in the advancement of the field. These issues do not seem to originate from any sense of uncertainty towards the unknown, but they certainly can indulge pseudo consequences before experiences could be time-tested. On the other hand, sometimes a hype is created by the people involved in so called *success* to generate an artificial scenario, which in the long run is realized as overindulgence and also becomes severely counter-productive because the chariot of expectation fly higher up in the sky! The current state of nanotechnology apparently looks to have gotten influenced from both ways. Such a situation, if appears confusing, obviously has an unforeseen cause inscribed within the collective operational modality of the process of conversion of scientific knowledge into commodity. Responsibility lies more on the scientific community in their evaluation and judgment on *where do they stand at present* in order to transmit *how much* in *what form* out from the laboratory, rather than on the people who started with a clean intention of creating a new business; this person could also start with an objective made out of a misunderstanding of the consequences of an yet unde-cided scientific reality, underestimating the time frame of maturity or simply by dreaming on the possibility of a revolution! Therefore, it is required to reanalyze the failure stories of many startups and enthusiasts rather than highlighting only the tales of successes! Nevertheless, science for its journey depends on policy mak-ers who are accountable to answer to the common folk in plain words, that of course many times becomes difficult without glossy words of hype or cherishing a dream; reasons apart, people sitting long time within the boundaries of labora-tories mostly forget the language spoken at the grass root. It is indeed nontrivial! History apparently takes similar course but changed time also has new attire.

5.2. NANOMATERIAL RESEARCHERS

This is the largest domain of nanotechnology activities that began in the '80s. The field also has attained certain level of reliability in terms of sellable products also. From steel to textile, size dependent properties of all materials are actually being explored largely because of the fundamental realization that *small is differ-ent*. In a simple sense, it means that if we can achieve the capability of interpolat-ing the manufacturing ability down to nanometer scale, we might have significant benefits irrespective of what the material is!

But, before taking an account of the involvement of different investments in this domain of activity, it is worth to review once again the purpose of research on nanomaterials so that we might be focused towards the goal and link different endeavors to comprehend better. At present, the target goals of all the efforts in nanotechnology might be identified broadly to belong to the efforts of,

* Clean and renewable energy;
* Cost effective healthcare for all;
* and in addition, though little distant, a third could be imagined to be safeguarding our planet from extraterrestrial invasion.

True, these are in demand across the globe and fulfilling them quite obviously requires highly coordinated international efforts over a spectrum of different pro-fessions in society. Participation is required from not only the scientist, engineers, technocrats and biologists but also the politicians and law makers, the corporate world and above all, a highly informed public! A systematic buildup of highly coordinated international endeavor is the ultimate key to success in uplifting mankind through the benefits of the new era technology across the barriers of religions, creeds and national boundaries. It is not expected that it would happen always through master plans and short or long scale projects or programs. However, works will be continuing through sustained buildup of interests and visions of benefits, documented, published, produced, manufactured and interviewed, reviewed, re-reviewed, to consolidate knowledge, muster expertise and measure achievements in terms of progress towards the final goal. In fact, all sorts of efforts as a whole could be encompassed within the structure of a virtual *enterprise*; that might in future be culminated into an action-driven non-profitable body like the United Nations. This could be the most realistic approach to avert the grasp of plutocrats or produce a handful of new billionaires at the cost of many different millions of investments. Within this all out drive, the humble motive of this book is to contribute in creating an informed public as a work force who might connect between themselves through mutual interests and associate with the objectives to create a future better for our habitat.

However, obligations of the legal frames of different nations would be the greatest hurdle towards achieving a synergy among different domains of activities in terms of projects within a single *World Enterprise*. To enunciate, it would be pertinent to highlight the news clips of recent nanotech events and conferences. For example, a news clip in 2010 from NSTI via email (from: NSTI [*matthew@www-nsti.ccsend.com*]) says, *"University of South Carolina (USC) reaches an agreement with DuPont on RAFT technology"*—invented originally by CSIRO, Australia in partnership with DuPont. This, to some extent sounds like 'US President has signed an agreement with the President of Russia!' Repeat of such acts and even the more vociferous ones in continuation, certainly cannot build any practical foundation for an international enterprise. Such acts would rather inculcate a type of paperwork to remain sealed within red-tape category for any practical purpose! At the most, this could initiate a situation like the existing process of paper-peace exercised within nations that is always there but not really without new anxiety on a daily basis to serve the purpose of augmenting the military budget! Therefore, a new type of internationally agreed trading practice should be evolved for knowledge products that could take care the profits by default within the limitations of a time bound objective, at least for next 100 years so that, for example, for an invention at an Australian company, a person at Oslo University wouldn't have to sign an agreement for furthering another new idea over that invention! The attitude of the corporate house has to change for a larger interest of humanity. Business has to evolve a new modality with more human face for a better tomorrow on our planet!

5.3. FROM RESEARCH TO APPLICATION

At this point it is important to look how research in fundamental science is related to possible applications culminating into achievement of major nanotechnology goals. This correlation is vital to visualize *what we are going to achieve as improved products for society* through *what we are striving to accomplish today inside laboratories.* James Heath, E. W. Gillone Prof. of Chemistry at Cal Tech, in his introductory remark in a recently published collection of reviews on nanoscience and nanotechnology, has illustrated a chart in this regard that assists to trace this interconnection. It would be most pertinent to highlight and adopt his classification for academic reason as it best manifests the interwoven threads of research, technology and its broader applications. However, the scheme is redrawn here in slightly different fashion. Fig. 6 shows the drawing. Let this apparently hammock-like construct be referred as *Nano-Net.* The right-side shows six terms that James Heath defined as *nanotechnology variables.* These are essentially the parameters that qualify and differentiate nanotechnology from the technology that we are now accustomed to. In our traditional practice in technology, we are not alert to any sort of dependence on size, shape or surface-to-volume ratio of the material used. It is meaningless to ask these things for common bulk materials, e.g. to make a railway track we do not refer to size and shape of the chunk of steel that it would be made from. The mechanical property of steel does not change with the size or shape or surface-to-volume ratio etc. and to be precise, these parameters do not carry any sense whatsoever in the macroscopic domain. Actually, these parameters are exclusive only to the scale of nanotechnology. They are variables because their values can be manipulated to modulate the properties of materials. In nanotechnology, we need to acquire the expertise to vary them in order to derive the best interesting material properties that have not been possible within the range of conventional practice of technology!

Though interrelated, it is preferable to consider size and shape as separate variables because any size might occur in different shapes or any shape often qualifies to various sizes. Small bimetallic clusters or nanoparticles have good examples of such typical variations. Among these 6 variables, 4 are essentially independent. However, super-structure and surface structure are dependent or composite variables made out of independent variables. The curved arrows on right-hand under the column *variables* shows this dependence more specifically where *super-structure* is shown to have dependence on size, shape and surface-to-volume ratio and the *surface-structure* is composed of size, shape and composition heterogeneity.

Even though size and shape are imaginable qualities, because of change in length scale they are sometimes quantified in nonconventional ways, e.g. size is quite often referred by the number of atoms or molecules contained in a nanoparticle when they are very small because the number is countable. At this size, in fact, every entity differs from each other. Whereby, an entity having 11 atoms may differ entirely from one having 10 atoms because they might differ entirely in terms of the arrangement of atoms with respect to each other in their respective structures and, therefore, it is a non-scalable regime. Similarly, shape is also manifested by particular structural arrangement that binds the constituent atoms (or molecules) one another to consolidate them as a single entity. In a similar way, surface-to volume ratio is something that becomes relevant only when we go to smaller sizes. For materials of our everyday experience, it is actually the volume that we talk about because the amount of surface is much less compared to volume and hence becomes insignificant for applications unless there is any special situation. But, if we want to construct a small 3-dimensional structure out of, for example, tiny spherical balls, it reveals that most of the balls would essentially lie on top, i.e. on the surface. For example, the smallest icosahedra structure has 12 vertices. Such a structure can be realized with 13 balls where each one sits at the vertices and the last one sits at the center. Observing closely, it would reveal that such a structure has a whole lot of surface compared to the amount of volume it contains. Though it is not particularly essential to know, how exactly an icosahedra looks like, it is mandatory to realize the fact that an apparently overlooked quantity becomes important as parameter when we go to smaller sizes. This parameter can be varied and even utilized to qualify finite small entities. For name sake, it is, *surface-to-volume ratio* or may be referred as *S/V ratio*. In analogy to biological names, one might use the variables as prefix or suffix as identity, e.g. $^{Ico}Na_{1n}$ meaning the 1 nanometer size sodium nanoparticle having icosahedra shape.

Therefore the parameters on the right hand side of Fig. 6 set an identity to classify a nanomaterial or nanoparticle or nanostructure. In fact, the properties of nanoentities are decided by these parameters that in turn certify the end-product in terms of applications. As a matter of fact, we need to break our brain to play around with these parameters in order to tune the properties of materials that are placed along the horizontal scale of the figure; they manifest the qualities in materials. The primary objective of playing around with variables is to achieve better and better or novel properties or novel values for the properties. This is where the bulk of research works reside that ultimately decides the realm of applications for benefits, which are placed on the left side of the diagram. We need our final steps onto the things called *applications* which endorse the level of comfort through the advances in our collective belonging and cohabitation.

The hammock is a network to represent different links within the world of *nano* and is shown with sets of colored arrows. It is necessary to follow where the different arrows terminate. Characteristic of that particular quality should be possible to be manipulated by those parameters from where each individual arrow originate. For example, each property can be manipulated by a set of variables or a combination of them. Likewise, one application can result from one or a set or a combination of different properties. Such specific dependences might be observed following the arrows of each individual colors. It immediately turns out then how immense is the possibility of expansion within this realm and how immense it might finally be! Such possibilities add up the extra dimension in nanotechnology hitherto seen absent in our experience of conventional technology.

Properties qualify and also classify materials. The knowledge of properties enables to decide how a material could be used for specific purposes. Measurement and determination of properties of materials had been the primary task at early stage of development in Solid State or Condensed Matter physics, which later gave rise to the more special branch called Materials Science. Knowing materials better builds the confidence to apply and utilize them better. After all, it could be worth remembering that the genesis of *Science* is from the special human sensibility driven by necessity like the realization, *putting an iron ring over the wheels of a horse cart makes the wheel stronger and last longer than making a complete iron wheel that makes it heavier.* Hence, the inner quest in *science* has all through been trying and finding an application of the knowledge acquired; thus testifying the abilities till satisfaction and therefore, it is never complete!

The six words on the bottom horizontal line in *Nano-Net* are the primary properties which we largely use to segment the entire subject of Condensed Matter. Any application, be it an aircraft or a PC-mouse, could be correlated to a material property or mostly a combination of properties. Though it is out of the scope of this book to elaborate extensively on what these different properties actually are, in a common language it could be depicted as if you have your solar cell device working because of the material that is capable of producing electricity out of light; you have your food items stored for future use in a refrigerator because of the thermodynamic property a particular gas; you have your rooms enlightened because of the electrical properties of the wires used to carry electricity from your power provider; you see things in your cell phone display because of the thermal-photonic properties of the liquid crystals therein etc.

Innumerable examples could be sighted from whatever you associate with in your daily course of life. Somehow or other, at each moment of your modern life, one or the other material property is actually helping you out! Therefore, it is extremely necessary to realize that the entire volume of things we use in modern civilization is basically an outcome of our knowledge and experience of properties; that's indeed the *bulk or body* which empowers our motion, e.g. applications, utilities, gizmos, conveyance and conveniences!

As James Heath proposed, the six applications listed on the left side of *Nano-Net* can be broadly grouped into two large areas viz. *Clean Energy* and *Health Care*. The first two applications from top on this list are actually meant to achieve more easy, improved and affordable healthcare system. The four applications down the list are meant for the production of clean energy. In his words, "*These are representative of nanotechnologies only in that they span the range from existing commercial products to avenues that are still evolving at the basic science level.*" This statement indicates a course of journey for the benefit of mankind, in applications derived out of the studies of properties in terms of nanotechnology variables. In order to have a projection of benefits from nanotechnology in clean energy and healthcare sectors, it is worthwhile to reassess where we stand globally today.

5.3.1. Clean Energy

Our globe is still in short supply of energy, not quite because we are not able to produce more, but because we still lack in the technology to store and convert efficiently from the available abundant sources in *nature*. If we would have the absolute capacity to convert and store energy, then even our everyday supply from *nature* would take us through a long time in future. We still burn fuels available inside earth and produce the harmful carbon oxides (both mono and di) to transport us from one place to the other whereas aplenty of sunlight is just wasted. Imagine if you could just place your car in the sun for some time and zoom out for the day! Our long term sustainability requires renewable clean energy without polluting the environment and without creating any waste or debris filling the earth or ocean. We recognize but still fail to materialize that the greatest boon on planet Earth is the huge natural abundance of energy from the sun in the sky and water in the oceans. All the energy that we need to survive is stored there! In fact, we inherited two fundamental things from *nature*! Matter, in the form of Earth, and atmosphere; Energy, created in the cosmos. We receive a part of energy in pure form. In actuality, the livings are essentially composed of a judicious combination of matter and energy. We know that matter and energy are specifically related and quantifiable. To date, this discovery of Albert Einstein, i.e. $E=mc^2$, is the single most important realization of human race for all centuries. In the process of evolution of human race, this is the second most important learning after human beings learned to make fire in prehistoric times! These learning are the biggest and most fundamental milestones in evolution. Further, we have also learned the ways and means to utilize the matter-energy relation and how to convert some amount of matter into energy (either in controlled or non-controlled way). However, we are yet to learn utilizing the equation in the forward direction, i.e. to create matter by spending energy, except producing some fundamental particles as the derivative of nuclear reactions conducted in large particle accelerators. The factor c^2 is so large that it is practically impossible to accumulate energy in order to create matter!

Neither do we know how to efficiently convert the plentiful supply of energy received at every moment from the cosmos into a form that is useful for our purpose, nor are we capable to store it in any other form. We still survive on our ancestor's discovery and burn things available to help us, may be in just a sophisticated way, i.e. do not burn the wood logs for all purposes but definitely burn coal or gas to make electricity and then use it to cook food! Latest in this score is the knowhow of burning matter through fission, also proving that we have become quite an expert and specialized in the means of burning things! In fact, all burning is associated with some form of pollution, either immediate or in the long run, and it is bound to impose some sort of disturbance in the environment. We need now to master a technology to efficiently convert the available energy in *nature* and benefit from what is already stored for us besides at the same time, inventing and evolving more and more precise devices, equipment, gizmos and applications that use less and less energy. Conceivably, it is not wise to copy the process that exists within the sun, rather it is useful to benefit from what it supplies as a product of that process being stranded at a reasonably distant location. That is the significance of the sun being located at a correct distance from our planet and this is what actually makes our existence unique in this solar system, if not in our galaxy or in the whole universe! It may be worthwhile to recollect that the total amount of energy received on earth from the sun per day is 10,000 times more than the total amount of energy spent daily all over the world! Quite obviously, we are not yet smart enough to utilize what is given to us naturally and on that token, we are still not accountable to *nature*!

Thermochemical fuel and thermoelectric materials are applications wherein the sun is used as a source of heat to generate power. Large, highly reflecting mirrors, referred as thermal harvesters, are used for converging heat waves from the sun's radiation onto collectors that capture and transfer the heat to surrounding water, producing steam that runs the turbine to generate electricity. In principle, the power generation modality is similar to that of normal nuclear power reactors, viz. boiling water reactor or BWR. The difference is how heat is generated. The heat produced in a controlled chain reaction at the reactor core is used there to produce steam to drive the turbine. Basically it is transformation of heat energy into electric energy in both the cases. However, the former is using a natural resource and the latter is from a man-made source. The heat production process in the former is actually nuclear in origin, mainly nuclear fusion reaction at the surface of the sun, which is happening at a distant location so that we on earth are not affected by any kind of debris whatsoever! In both the cases, eventually matter is converted into energy, but here we work with the emanated energy in a clean form to start with and convert it into our useful form, viz. electricity. In the later process, we burn matter in man-made artificial hearths (or machines) and along with the heat, we also produce high energy particles, radioactive materials and radiation, which are all potentially hazardous for all living beings. In fact, heat is kind of a by-product in a nuclear reactor and has not been the primary motive. Therefore, power from a nuclear reactor is essentially auxiliary. Moreover, the world is still not in total agreement on how to abolish or really what to do with a dead reactor. Over years and years, all nuclear corpses will stand on earth as zones of hazards prone to accidental disasters that the world has already witnessed more than once! True, today's technology of thermochemical and thermoelectric power generation is not totally efficient compared to power generation in nuclear reactors in terms of total power delivery, but in terms of infrastructural, maintenance and safeguarding expenses, investments are much smaller in scale than that of a nuclear power reactor. Therefore, the expenses to commission a single nuclear establishment could be compensated by installing many small thermal solar units. After all, the sun is there all over the earth's surface at some point of time except on poles which we should solve differently. To quote James Heath in this regard from his essay, *"Electricity produced with this approach can cost as little as 10 to 15 cents per kilowatt hour."* This is comparable to the current cost of nuclear power which is the cheapest among all different kinds of power generation methods.

Research in nanomaterials has a very promising role in order to make this process more efficient by improving the capacity of capturing and storing solar heat more proficiently. Actually, the collectors will be improved in their capacity to capture more heat and eventually the entire process can be made comparable to the cost of nuclear power production. To correlate, let us answer the question, what actually is the thread here from variables to properties to application? The total heat capture depends on the amount of surface area of the collectors exposed to the focused heat from the reflectors. If the collector is made out of an assembly of nanoparticles or nanostructured materials, the amount of surface area increases by several orders of magnitude within the same overall dimension of the device because nanoparticles have higher surface-to-volume ratio and the properties could be manipulated through the control of size and shape. This enhances the heat absorption property resulting consecutively into much larger turnover in application, i.e. production of electric power. Incidentally, it might also be worth to mention here that considerable amount of insight is developing for new design for natural power harvesting through the research in bio-mimicry and together with the potential of nanostructured materials, it should be possible to harness a new dimension in the utilization of natural resources of energy. True, we have to still wait to replace nuclear energy that is the cheapest source of intense large amount of power required for the industry. Nevertheless, efforts on other secure and inherently safe avenues could surely limit us from mindless extension of nuclear power units to live the life within potential threats of accidental disasters. We should remember, after all, the experience of Chernobyl could not prevent that of Fukushima!

In photovoltaics and photochemical materials, the light from the sun acts as the source of energy. The task here is to convert light energy into electric power. As such, for quite a long time now the world is familiar with electronic gadgets that work with light. For example, small devices like calculators, wrist watches etc. and many others. In fact, solar power is practically synonymous to the use of sunlight for electric power. Whenever we see the arrays of black panels fitted on rooftops or over water heating systems or a layout of such things spread over an open area, we know that it is for solar power, either in lighting up a building, or heating water or running a cooking range etc. However, these are all small applications in which the power required to drive the system is not large. In fact, power using sunlight still falls short to run industry or even a locomotive. There are indeed major challenges to overcome at fundamental level in obtaining a large quantity of power using the photons (i.e. quantum particles that light is composed of) directly.

It will be pertinent here to reinstate the nomenclature of different phenomenon that happen on the interaction of light with different materials in the perspective of electricity generation. However, the exotic phenomena that happens on the interaction of intense laser light is excluded because that is a different story and yet a burning subject of academic research only. Further, when it is referred about electricity generation using light, it is to be followed that light here is essentially considered through quantum mechanical description. Light is composed of the energy particles, viz. photons, that has zero mass and energy $E = hv$ and momentum hv/c where h is an universal constant proposed by Max Planck and is known after his name as Planck's constant and v is the frequency of photon vibration. Imagine these photons as tiny packets of energy, known as quanta that cannot be used partially; either to be absorbed or rejected in discrete numbers. Therefore, when light falls on material, in the micro level it is essentially an interaction between photons and electrons within the material. Depending on the material used, the responses vary. In the right kind of material when the photon has energy above certain threshold, electrons eject out of the material and they are referred as photoelectrons. If there are suitable means to channelize these electrons, electric current could be produced. This is photoelectric effect. However, as and when an electron goes out of the material system, a void is created within it. It is referred as the creation of a *hole* and to have the effective charge neutrality, this is effectively the image of electron with opposite charge. Thus, a voltage difference or EMF can be generated. Hence, the interaction of the right kind of photon generates an electron-hole pair that could be transformed into power. More effective is when a common junction of two dissimilar materials is illuminated; a larger voltage is produced. Here, we refer this as photovoltaic effect. Several important applications are found to employ this, e.g. photocell that utilizes the voltage developed across a metal-semiconductor junction as a measure of the intensity of the incident light; solar cell that uses the photo-voltage developed across a *p-n* semiconductor junction as a source of power. In fact, an electric field is established at the junction of two dissimilar materials and the contact potential at the interface is the source of this electric field driving the photocurrent. In photoelectrochemical fuels, the separated electron-hole pairs are used to drive multi-electron electrochemical processes. Reduction of carbon dioxide and splitting of water are the most common ones. Since in all phenomena, surface and interface essentially plays the key role, use of high quality nanoparticle assembled materials and tuning of the surface-to-volume ratio parameter should lead to more effective surface and better quality interface. It should be recalled that homogeneity of the interface is extremely important for diffusion of electrons because a system of two dissimilar materials joined together attains new thermodynamic equilibrium and the Fermi level in both becomes equal. However, in view of such huge applications, even an incremental development in this area has a hope of large economic benefit.

It could be worth knowing here the limitations and obstacles in obtaining 100% efficiency in photo-cell applications. Theoretically, the sunlight conversion efficiency of solar cell can be as much as 90%. However, this cannot be adhered to due to other fundamental and practical limitations. Photons with energy less than the *p-n* junction band-gap cannot be utilized. Further, the electron-hole pair with energy far greater than the band-gap relaxes very quickly to the band edges, releasing the excess energy by heating up the cell itself. These facts limit the maximum efficiency that is at the best around 30% only. These facts impose the limitation that, to date, solar cells are specific and there is none that covers the entire electromagnetic spectrum. Overcoming this *Shockley limit* is by itself a topic of current research in nanotechnology.

Shockley-Queisser (SQ) or the Detailed-Balance limit is in fact a fundamental rule on the energy conversion efficiency of solar cells that originates from the 2nd law of thermodynamics and is one of the most important theoretical insights for solar power generation. For single (*p-n*) junction photovoltaic cells, the maximum achievable efficiency is calculated to be 33%. The original value given by Shockley and Queisser for silicon solar cell was 30% and the best lab achieved data till date is 25% obtained recently at University of New South Wales, Australia. At module level, Sun Power Pvt. Ltd. has reported in March 2012 to have achieved 20% efficiency.

The trick of overcoming the SQ limit has taken research in

1. Finding new materials and
2. Finding innovative carrier generation process as well.

As a material, Gallium-Arsenide (GaAs) can generate a decent 25%; however, none other material in single junction mode could actually compete silicon which on the other hand is a lot easier to handle than GaAs. Formation of more than one junction material is quickly visualized as one promising approach and multilayer (2-3 layers) cells of different *p* and *n* type semiconductors, e.g. GaAs, InGaAs, AlGaAs, GaInP$_2$, AlInP$_2$ etc. are developed resulting into a conversion efficiency as high as 38%. However, these stacked cells have several fundamental drawbacks in scaling them up to commercial grade power production and research efforts are continuing to overcome various issues. The trick of the trade meanwhile has shifted towards other directions, including finding innovative applications of the novel properties of suitable nanostructures to enhance the conversion efficiency.

Theoretically, it seems that 47% of solar energy is converted to heat, 18% passes through the cell and 2% is lost through recombination to give an ideal 33%

to convert to electricity in which 6 to 9% is actually reflected back from the cell in a practical device! A closer inspection of the electromagnetic spectrum shows that infrared, microwave and radio have insufficient energy whereas X-rays and γ-rays have too high energy to get absorbed at all. It is only a tiny band slightly above the band-gap of the material that involves in charge carrier generation in which the UV photons generate charge carriers with much excess energy that relaxes at the band edge and are not useful to convert to electricity. These high energy electrons are referred as *hot-electrons* which generate excessive heat over the duration of operation and diminish the efficiency over a period of time. Three broad fundamental approaches could be identified to circumvent these issues:

* Generating more than one $(e^- - h^+)$ pair per photon that has energy higher than the band gap, like the inverse Auger process. Certain semi-conducting nanoparticles or Quantum Dots are speculated to have good promise in this regard.
* Inciting a two photon absorption process by introducing an intermediate band to exploit the photons that have energy less than the band gap. These intermediate band semiconductors are not easy to manufacture and a good one is yet to be tested.
* Separating the charge carrier generation and its transport within two different materials juxtaposed together in order to avoid the recombination and heat generation slowdown in performance. Dye Sensitized Solar Cells or DSSC is a good example of the phenomenon but yet limited to 12% conversion efficiency.

The research work on all these aspects actually constitutes the objective of *3rd Generation solar cells*. A detailed account of research findings on all these current directions is out of the scope of this book; however, interested readers may consult the bibliography for further reading.

5.3.2. Spend Less to Get More

Nevertheless, this is just one side of the story that talks about production of energy, amply and more sensibly. The other part in fact lies in our everyday stories of consumption. The hunger for energy will always increase in an ever expanding civilization; more so, because electricity is taken as an index of progress in society! How to put a limit to it is a different question, but no amount of production and conversion would suffice until we at the same time realize to consume less and work to implement it as a primary attitude. This is where the games in *nano* again have a strong role to play by changing things to work with much less input of energy and thus revolutionize everything to a different echelon of (lower) power consumption without any loss in efficiency. This of course is not simple and cannot be achieved only through modifications! Rather, it demands a different approach at the fundamental level along with readjustments in our practices, habits and attitude. Let us cite a small example. We have come far off from the bulb that Edison gave us. Think about the new energy saver bulbs (CFL) available at market today! A 10 watt bulb has the same luminescence of a previous 60 watt one! The advantage is that it produces much less heat because of the use of much less current. Adaptation is that we got to be used to cool white light, a feeling different from before. The new application of LEDs (Light Emitting Diodes) in this direction is remarkable! LEDs exist for a very long time now in electronic appliances. Very recently, it is being used for lighting the flat panel LCD displays of TV. With the invention of intense blue and white LEDs, home-lighting systems have flooded the household market. The advantage here is that one can have very localized but intense lighting. In fact, a NASA night picture shows how much light goes out to the sky at night from the entire globe so that the rich and the poor parts on the planet could be easily identified even from the outer space! This entire amount of light energy does not serve any purpose and is practically wasted which means that we are wasting a lot of our precious energy just because of application of inappropriate gadgets! Lighting should actually be localized as and when it is required. Applications of LEDs are very promising in this purpose. Even for an aesthetic and scientific lighting design (like the proposed South Korean skyscraper), along with the use of more daylight in the buildings, combination of localized lighting with shadows proves to be more soothing, if not only a way better for mental health! A good example in this regard is the British Science Museum building in London. SEC Solar Energy Centre (www.secbattery.com) has a remarkable achievement that should be mentioned in this regard. Their commercial LED light bulbs use only 2 watt power to deliver the luminescence of a 40 watt incandescent lamp. Energy saving by a factor of 20 right away! Moreover, with their solar rechargeable batteries this light can glow fully for a continuous period of 7.5 hours. Using ultra efficient LEDs that are more than 20 times competent compared to incandescent lamps and 5 times more than fluorescent lamps, these lights use only a maximum of 4 watt input power for a 480 lumen light (~ 80 watt bulb). Such small energy requirement can even be backed up with solar power batteries for more than 10 hours of operation. At this continuous operation rating these LEDs have a lifetime of 2.3 years, much better than even the fluorescent lamps! Such products should be subsidized by governments of all countries and they should be facilitated for mass production all across the globe so that very soon these lights take over the profuse use of fluorescent lamps all over our planet. Calculate, how much savings we can do to our energy resources just by ushering in a complete replacement? Products like this give opportunity to decouple entire illumination consumption from the primary power production grid in a city vis-à-vis a country and the whole world. It should be effective to generate an economic benefit also. However, we need a change in our attitude to orient our endeavors and we also need to reorient our psyche that the more our national boundary is visible from space should not mean the more improved we are as a nation!

It not only requires research in bulb technology but also in battery technology which unfortunately is still not in a highly improved state. We are yet to have high-power, long-lifetime and robust batteries that might work for weeks and months at a stretch without recharging. Neither do we have very affordable high energy density battery backup systems for our power grids except those that are supported by pumped-hydro and compressed-air conversion backups. Battery banks installed at far-north and far-south icy zones can pull up only for a very limited period of time though the cost is very high. Nevertheless, fundamentally the knowledge behind all these progress boils down to conducting more research on materials, i.e. finding new materials or incorporating new properties into existing materials that will be useful for these applications. Lots of researches are continuing on graphite intercalation systems for advanced battery applications but there are still limitations. Nanotube, the rolled over graphitic sheet and its intercalation properties, graphene, the infinite single graphitic sheet of carbon atoms, are gaining importance to be explored for more innovative power-storage applications. Otherwise, as new materials, titanium dioxide nanotubes also are revealing more intriguing results to be exploited into innovative applications. We still have scope to imagine materials, *lighter than cotton but stronger than steel*! Likewise, in machineries, we should imagine, *lesser in fuel but greater in efficiency*! This perception of *more-for-less* improves the production coefficient. Motors and engines should be far superior so that it consumes much less to produce the same amount of work output that we have now whether it be a lathe-machine to make metal components or a locomotive to mass-transport over long distances. It should be observed that when things change at fundamental level with this objective, the usual sense of *loss-factor* in the process no longer persists or becomes redefined. Therefore, the knowledge and expertise that we have should work as the basis for an all-out new endeavor. The domain of technology at nanometer dimension heralds a huge promise to usher in a paradigm shift in this direction.

5.3.3. Healthcare

So far, the applications in energy sector in terms of how to harness clean energy and the efficient use of energy has been discussed. The next big target application of nanotechnology is in healthcare. Like the agenda *clean-energy* in the energy sector here, the agenda is *amenable universal healthcare* for everybody across all the nations as a fundamental right to humanity. It is true that good nutrition and healthcare for all at every corner of our globe is not yet there. Every year many hunger deaths are recorded all over the developing world, especially in poor African nations and even in India! Modern medical science has of course attained many successes but is yet to become affordable and thus remains out of reach to millions in the developing world. Sophistication brought in an additional price tag: reduction in rate of mortality has come at the cost of heavy dependence on machineries and equipment; there is no true knowledge on the exact quantity of any drug required for a treatment; neither the actual origin of an ailment down to cell or molecular level functionality is still known. These are large open issues common to the entire globe and surpass all political boundaries. A lot are there to be done! Therefore it is most welcome if any new technology on the horizon can ring a bell of new hope. This is where lies the importance to be more intense in research towards humanity, towards a radicalization through government and political collaborations across nations, more through a genuine humane face in global business, going beyond charity, but incorporating active participation.

Nanotechnology has raised a lot of hope for sensing a disease at a different level of precision that it would be kind of independent on what a patient transpires about his feelings when caught in a disease. Together with the expertise in biotechnology, it shows possibilities of access at genetic level. In fact, all domains of medical science are carrying out extensive research to fit to certain goals. Reports and papers are pouring out almost on an everyday basis. Certain core areas that have lots of promises could be identified as *Drug-delivery*, *Pre-symptomatic-treatment*, use of *Biomarkers* and *Regulatory-sensors* etc.

In today's medicine, the strategy is to first detect a disease and treat it after, sometimes at a later stage or for a prolonged period. This practice is referred as *reactive medicine*. Through implementation of nanotechnology inventions, the objective in medicine practice would shift more specifically towards *preventive medicine* and we know, *prevention is better than cure.* In fact at the moment, only in very limited cases we do have the preventive practice, which is limited mostly to child healthcare. The difficulty in initiating a preventive practice is either it has to be a stereotypic problem or one has to first precisely know the root causes that are to be truncated or eliminated. It means, the entire process of medical practice should have to be proactive and better, personalized. Nevertheless, the approach should not sound like we all would become some sort of medical guinea pigs under constant vigil and monitoring and lose our freedom in the process! Certain parameters of our physique and somatic-sensitivity data would only be stored and monitored in order to predict and control the occurrence of specific ailments. Employing extremely tiny (*nano*) chips mounted within our tissue, essential body parameters could be continuously monitored for a particular disease. Having the gene-map data of an individual done right at birth, it might even be possible to steer a body from getting infected and eventually cured by applying specific inter-active drugs delivered at precise locations. In other words, there would be limited scope of diagnosis in the conventional sense that involves an element of guess. The proposed new approach is contrary to the conventional one which depends on the experience of the doctor having dealt with a similar situation previously. In the new approach, we are interested in precision and application of personalized med-ication even for the same disease but in two different bodies. In fact, disease, i.e. the body being not at ease, even if it is caused by the same reason, in two separate bodies, truly, they should be considered as two different diseases! Because of the fact that medicine would be precise, it would obviously also be very little in dose. Further, because the entire process shifts towards prediction-assisted prevention, chances of surgery would reduce drastically that it might not be necessary at all except in unavoidable circumstances. As such, surgery is an emergency relief only and not really a treatment! To some extent it is indeed primitive in approach but through technological sophistication it could ease and become successful vis-à-vis powerful as well. On second thought, the entire process of undergoing a surgery is like declaring war with the human body which is a most-sophisticated, com-plete, self-sustained and delicate factory. It is like dishonoring the greatest bless-ing that biological systems self-recover unlike the man-made machineries which need to be remade in order to recover!

In fact, the power of preventive medicine is already established and we under-stand its potential. Take for example, the case of polio. It is eradicated from the developed nations and almost also from the developing countries. The expense of en-mass polio immunization around the world is far too less compared to the cost of treatment for even 10% of the total number of people infected. Further, it is not actually guaranteed whether the patient would really recover completely through treatment once infected! Disease gets cured but in many cases it leaves a signature behind. Another recent example in regard to preventive practice could be related to the story of bird-flu H1N1 case that has occurred as an epidemic in certain places. It is soon realized that prior measures in the form of prevention has to be in force because it is extremely difficult to manage once the virus breaks in! The deadly virus spreads even faster than wildfire and the additional stress onto our resources and society becomes a heavy penalty! However, our medical science has not achieved a level where prevention can be undertaken for all different ailments; besides, it depends on several factors that at times stand as stumbling blocks to research. For many deadly fevers, we do have certain preventive measures with limited success, but not for a heart-attack, not for a cerebral stroke, neither for lifestyle diseases like diabetes, high blood pressure etc. Moreover, it is still a matter of debate particularly to decide how far these are really lifestyle borne!

Nevertheless, science is addressing the issue at very fundamental level and the world is soon going to witness a solution that is universal in approach and carries a lot of hope. Maybe it will happen through genome research or the study of genes or by deciphering the genetic information of every individual and keeping it as database to use for diagnosis and preventive medicine.

5.3.3.1. NANO AND GENO

Gene is a segment of DNA molecule. But any segment of the DNA molecule does not qualify to be called a gene. As characteristic, it should contain the information required for the synthesis of a functional biological product. Information is coded in genes that describes everything of an individual starting from his/her anatomy to health and mind to behavior. Analysis could also predict the future possibility of his or her ailments. To describe how information is coded within genes it might be easier to follow through the analogy of computer language.

We have languages that we use to communicate in which the fundamental ingredients are alphabets that make words and words are used to compose sentences that follow certain syntaxes which are (set as) rules created through convention and verbal practice over ages in different civilizations across the world; sometimes in slightly different ways and thus giving birth to different languages. However, the fundamental structural construct essentially has been the same except in some languages, e.g. Chinese, Japanese or especially those derived from ancient *Mandarin*. These languages have constructed words as fundamental units rather than alphabets making them much richer. In fact, a close analogy could be drawn here in the context of molecules and nanostructures wherein alphabets are compared to atoms which are as many as the number of elements in the periodic table. The world of words, e.g. *molecule* and *nanostructure* are far richer and larger! Like a dictionary of words is never complete, a dictionary of molecules is incomplete as well! Nanoclusters and nanostructures are creating an even larger new dictionary.

Apart from the language of our everyday use, in modern civilization we discovered a second language which is written using numbers that basically stands to express the relations and operations between numbers; in general, we call it mathematics. We use both letter and number languages in our everyday life retrospectively! Now, when it comes to instruct a machine to do our job neatly, because we know that it can handle the job faster and precisely, sans the factor of human error, we needed to create a script for the machine because it only can respond to two of its physical state of switching the logic gate, i.e. ON or OFF (fundamental electronics). Representing the two situations by 1 or 0, a language is created wherein our communication is coded in terms of combinations of 1's and 0's. Simplistically, we refer this as *binary*. But then, this script turns out to be enormously long and tiring to practice even for a simple communication. Therefore, a third interfacing forum was required that is a judicious mixture of our common and mathematical languages and the machine is instructed to decode and interpret it as binary in order to carry out the task asked for. This is called the programming language which for a layman is a kind of jargon, but for a programmer it is a specific set of instructions following certain syntax. But, even for the programmer, if a print of the whole binary is given and asked to carry out a reverse operation in order to produce understandable sentences for common man, ooops! It is unreasonable, unfathomable and a mad task. However, there exactly we are when we talk about decoding the genetic information! Over and above, there not only are 0's and 1's, but combinations of four numbers, e.g. 1, 2, 3, 4 corresponding to the four bases viz., A (Adenine), G (Guanine), C (Cytosine) and T (Thiamine) in DNA or U (Uracil) in RNA. Hence, what we have as a script (a gene) is a sequence of these four alphabets as representative of life's chemical alphabet. The study of genome is indeed a study of the sequence of these four letters in all genes and decoding the information hidden therein. A genome might be looked like a book written without capitalization, punctuation between words and paragraphs. Moreover, nonsense words and alphabets spread all across between and even within words. A phrase from this book might look like this:

asfgh**walking**klmnop**down**ttyes**the**tyesgh**street**kmldgsahgdf**dog**ghyyeicsujkhgd**den**lyyyglk**jumped**on**a**p**post**man**kkjhgh**hiding**in**mmndsthe**bush**opopytatbbbq**road**asgs**ide**

The bold faced portions only communicate something that we are able to understand! However, this example is written with a language familiar to us so that we can understand the meaning in the expression of the bold faced alphabets. Actually, *genome* is in an unfamiliar language, pattern of expression, difficult to derive any syntax and contain repetitive portions that make the task of interpretation extremely difficult. There might be an underlying mathematical rule governing the entire coding process, but yet to be discovered! Nevertheless, the genomic sequencing task that endows knowing the details of the exact order of 3 billion bases in all the 24 chromosomes in the nucleus of a human cell, is daunting, but unavoidable. The complete information with these 3 billion letters makes the genome book, different for every individual, nonetheless to be written only once and for all. This task is achieved for the first time in 2003 under the Human Genome Project. Achieving this goal has helped reveal the estimated 20,000-25,000 human genes within our DNA as well as the regions controlling them. The resulting DNA sequence maps are being used by 21st century scientists to explore human biology and other complex phenomena. With continual improvements of sequencing speed, reliability and expenses, the average time has remarkably reduced nowadays to about a year using completely automated processes. Worth to remind that first human genome work began in 1990 and ended in 2003 under International Human Genome Project! It shows a great hope that in about few years, genome sequencing would reach an affordable range of time and cost. This is being possible because of automated machines, supercomputers and to a large extent it being a repetitive measurement. Solving this challenge once more depicts that achievements in science is in fact associative and also amenable to the technological advances in different other fields in consonance.

However, sequencing the genome doesn't lead promptly to the genetic secrets of a species. Serious tasks remain to be carried out in order to translate those strings of letters into understanding the functions of the genome, e.g. how different genes are related in the genome, and how various parts of the genome are coordinated. This amounts to translating the meaning of the genome, i.e. how genes work together to direct the growth, development and maintenance of an entire organism. But genes account for less than 25 percent of the DNA in the genome, and therefore, knowing the entire genome sequence helps to study the parts of the genome outside the genes. This includes the regulatory regions that control how genes are turned on and off, as well as long stretches of "nonsense" or "junk" DNA (recall the apparently nonsensical portions of the analogy paragraph above) because it is not yet known whether they really have any function.

As creating the genome data becomes affordable, it should be made mandatory to have it recorded and stored for every individual and each newborn at birth. A new form of scientific birth chart hence could be introduced replacing the astrological (savage!) chart that is still prevailing within the value system of even the advanced societies! Perhaps this could initiate the building of a science based health prediction system. A yearly updateable chart or *Genoscope* could be initiated because people love and are used to seeing the horoscope, sometimes addictively that even Google also takes its fair share in the practice! Though scientifically there is no base to the horoscope, in order to satisfy the everlasting quest of people to have the mental pleasure of knowing about their personal future, it has a good business! Therefore, it might not be unreasonable to tap the potential of people's habit to consult the *Horo* and slowly turn the attitude towards *Geno*. Of course, the new scope wouldn't predict about individual wealth and relationship. Nevertheless, it might eventually usher in a peace of mind by asserting the old saying *Health is wealth*.

Patients can at least be given important information regarding the future of their health and physicians can become alert prior to the development of symptoms. As an outcome, preventive medicine would commence which is much less costly and free from the burden of hospitalization. Centers with large supercomputer systems can be established to work as data banks that would store the Genome data of each individual against the identity of their fingerprint and SIN (or Social Security) numbers so that even on accidents, doctors won't have to question the patient but only to retrieve the genome data! Hospitals can be centers for routine preventive advice and especially, for handling emergency situations. To elaborate a little further, let's examine cancer.

Today the normal course of cancer treatment is an initial judgment on symptoms, followed by diagnosis through physical inspection, some imaging technique or endoscopic procedure. It then follows through biopsy as confirmatory test. Then start the treatment, either by major surgery, followed by chemotherapy or radioactive radiation therapy. In many cases multiple treatments are suggested where one follows the other, in case the patient does not respond to drugs as anticipated. This entire approach is referred as the *Germ Theory of Disease*, by Dr. Danny Hills (founder of API, San Diego, CA, USA). As he propounded in a TED talk, the concept behind the conventional approach of treatment is essentially to look for a germ before treatment. Physicians are trained through this methodology that is highly successful for the infectious diseases. However, cancer is an unusual state where the affected cell systems do not perform its normal course of activities. It's an abnormality; neither caused by an external species nor as a result of any injury from external reason! It's kind of a broken state of functioning of a certain sector of the body and in general the cell nucleus is more responsible than the cytoplasm. As a result, the patient is in a broken physical state. There is no separate disease or it is misleading to say, *"His body is infected by cancer or he has cancer."* That sounds like an enemy got into the system or a sensation as if some kind of a deadly dragon has attacked the body! Actually, it is only a failure of the system, or a state that we should rather refer as *Cancering.* Changing the scenario into the form of a verb essentially changes the perspective as to how we should project it for treatment. It should also be reviewed how we categorize cancer in terms of the name of different body parts, e.g. breast cancer, blood cancer, lung cancer etc. In fact, even if it happens at the same organ, it is indeed different for two different individuals because the course of synthesis of proteins in two different bodies is different! Therefore, the treatment should actually be custom designed that is applicable only to a particular patient. With available genome data and advances in proteomics research, it would be possible to analyze the malfunctioning cells and, combined to in-site delivery of drugs achievable through nanotechnology, would lead to much specific cure. This would avert (1) trial and error of chemotherapy which sometimes acts like poison to the body if not effective for treatment and, (2) exposure of drugs to the healthy cells that we cannot avoid in today's inter-venous procedures.

5.3.3.2. THE PROTEO

Even though the genome is the blueprint of life, this data alone does not suffice to understand life's complicated processes, or the fine scale activities within the cells of our body. Like, only the blueprint of a building does not mean the building or makes a building, genome data alone does not actually empower us to decipher how the living system works! We need to know what and how different genes do a job. Which protein is created under certain situation and when? What we should benchmark as a normal healthy situation so that we can differentiate it from an abnormal or unhealthy situation? These are indeed intricate and in fact connectivity issues, within a single cell and between the cells as well. As Danny gives an analogy that genomic data is like having the ingredients in a food item that does neither qualify the food nor its taste; similarly we have to know which gene is responsible for which protein under what condition! Therefore *Proteomics,* the study of all proteins, i.e. (1) knowing the characteristics of individual proteins, (2) the sequence of amino acids within them and (3) the different transformations and changes that they undergo due to various cell processes etc. are all extremely important. It is a tedious, expensive and time consuming task to find out how and how many of the 24 different amino acids are stacked in a particular protein. However, with the possibility of rapid DNA sequencing, it has now become possible to know the full details of a protein much quickly. A protein sequence can be obtained from nucleic acid sequence that is also helpful to understand the connectivity between a genetic process and proteome. Also, modern sophisticated high-resolution mass spectrometric methods coupled to imaging tools have been devised to take snapshots or a protein map, for example, the entire proteome in a blood sample. The hard endeavor of Applied Proteomics Inc. is certainly appreciatory and promising to pave the way for an analysis of Cancering in cells. The answer to stop Cancering or cancer occurring in fact lies in our in depth of understanding on how different proteins are synthesized by various genes and when actually some genes are triggered on or off for such jobs. In this regard, it is worth it to remember here, Anna Barker, Director of Transformative Healthcare Networks, Arizona State University, who said, *"Understanding proteins would be the key to understand cancer."* Nevertheless, a detailed discussion on Proteomics is little involved and out of the scope of this book. To conclude the discussion, it is relevant to resonate once more with the observation of Dr. Danny Hills, *"Genomics shows us a list of the ingredients of the body—while Proteomics shows us what those ingredients produce. Understanding what's going on in your body at the protein level may lead to a new understanding of how cancer happens."*

5.3.3.3. BIOMARKERS

The details of the scientific methodology and technique of data analysis is actually not relevant for this book. Interested readers might anyway follow these intricacies through bibliography. It would however be incomplete not to mention the most genuine trick of the trade that is leading the way to transform the future of our medical practice from *diagnostic-predictive* to precise *probabilistic-preventive* regime! Hence, it is inculcating a paradigm shift of the way we treat patients. Biomarkers are the staff that essentially is making things possible! Its usefulness is so much evident now that the concept itself should be recognized as really a novel one. A biomarker is the short form of biological marker which normally is a substance used as an indicator of a biological situation, e.g. a normal biological process, pathogenic process, or pharmacologic response to a therapeutic application. Conceptually, it is not new and in fact, has been used in many fields of scientific studies. It has a wide range of application in medicine, e.g. biomarkers are introduced into a specific organ to examine the proper function or any specific health aspect. Rubidium chloride (radioactive) is well known as a biomarker in evaluating perfusion of heart muscle. However, certain selected and well known proteins are more specifically used for the job where a change in the state of that protein might indicate the risk or the stage of a disease. It might also be used to test the response after administering a drug or a specific course of treatment. In epidemiology and toxicology, external substances or their variants can be a biomarker. In cell biology, a biomarker is used to detect and isolate particular cell types. For example, protein Oct-4 is used as a biomarker to identify embryonic stem cells. Biomarkers in genetics are often a particular sequence of DNA identified to cause a disease or associated with it. Circulating tumor cells and micro RNAs are also used as biomarkers. If a complete assay of proteins in blood can be done, it could be possible to asses some blood-based biomarkers to provide pre-symptomatic information pertaining to the progress of cancer. It is then possible to administer therapies that would affect just the disease cells. The patient might then be monitored using those biomarkers that indicate the disease. Human blood proteome is very rich and blood-based proteins are very likely to provide major breakthrough in monitoring patient's disease and health when combined with applications, such as site-specific drug delivery and mountable tiny regulatory sensors.

5.4. Nano to Bio

A number of nanotechnology applications might be identified in biological sciences and such Nano-Bio areas of research are producing remarkable results every day. Some prominent examples:

* Drug-delivery
* Bio-labeling and bio-detection by in-vitro disease diagnostics
* Regulatory nano-sensors
* Nanotherapeutics and in-vivo imaging
* Regenerative medicine, external tissue products
* Internal tissue implants, xenotransplants
* Bone and dental replacements through biomimicry
* Miniature medical devices for improved endoscopic procedures and completely controlled robotic surgery
* Nanorobotic (nanobot) therapeutics,

In fact, progress in each domain very well depends on simultaneous advances in the other fields of science. For example, the use of tiny sensors depends on developments in polymers and fibers that also have applications in textile technology. Fabrication of tiny medical tools needs research on metals and mechanical properties of materials etc.

5.4.1. Drug Delivery

A substantial portion of the drugs that we take orally is actually removed by our liver and is not in use to serve any purpose. At the current situation of technology such wastage of drugs is unavoidable as the drug intake depends on the functioning of our digestive system. On the other hand, it is also not possible to administer all the drugs via injection or intravenous procedures. However, if we have a more effective method of delivering a drug only onto the diseased part of our body, such losses could have been averted. In other terms, it would have been a savings to the pharmaceutical industry. By itself, oral drugs hold a huge global market to the tune of approximately $35 billion USD with an average growth of more than 10% per annum.

It is difficult to acquire a globally updated and complete list of all the products because it augments every day and many country-specific local versions are not found recorded internationally! R&D in finding an efficient drug delivery procedure is therefore as important an effort as finding a more effective new drug. Research in this area has in reality had fallen short because there was no anticipation to create an effective gain until some vision came through nanotechnology. Because we can now deal with very small sizes and capable of building machineries at very small scale, we are able to devise non-perforating methods to deliver drugs in an ingenious way at a particular location within our body where it is specifically required! For example, the chemotherapy drug that travels with blood all through liver, heart and kidney, is always as a major threat for unaffected tissues. There are even examples of drugs which are never in complete use for therapy due to various other reasons. The solution to such issues lies in devising more potential delivery mechanism for various drugs. A number of different new ways are at advanced stage of research, e.g. smart pills, gene therapy (vaccine), drug-polymer complexes, use of oligonucleotides, light sensitive quantum dots etc. to name a few. It is not mandatory to know the details about each of these techniques, which are basically a judicious mix of products and procedures. Possibilities are there through nanotechnology developments to not only deliver the drugs right at the tissue where it is needed but also to monitor the course of its travel through the blood vessels and while in action within the tissue. Such happenings are not too futuristic but seem to be around the corner. For example, with the extension of understanding on how the functions of proteins depend on the cell pH, it would be possible to guide the action of a drug as to when it should be activated. The possibility of attaching a light sensitive quantum dot with the drug-polymer conjugate opens avenues for actually monitoring the course of action through a precise control of cell pH and hence transform the action like an on/off control switch. Such a situation is still in anticipation; nevertheless, should be adored like a projection!

5.4.2. In-Vitro Diagnostics and Bio-Labeling

In-vitro diagnostics are used as confirmatory test and are becoming more frequent practice because of near precision results. Lab-on-chip micro-technology has demonstrated promising prospects for in-vitro diagnostics. This technique is applied in many different diagnoses, e.g. food allergy, pollen allergy, cat allergy, diabetes, malignant hyperthermia etc. and even for evaluation of food hypersensitivity. In general, the approach is the application of a physical device or chemical method or physiochemical methodology as biosensor. Undoubtedly, these are new means in biosensor technology. Special devices have been designed for in-vitro monitoring of glucose and glycol-hemoglobin HbA_{1c}.

Semiconductor nanoparticles or quantum dots (QD) have been illustrated to function as an effective bio-label for in-vitro diagnostics. Compared to conventional use of molecular bio-labels, the additional advantage here is that a control might be achieved by regulating the physical parameters. It is also possible that different size QDs of same material might act as bio-labels of various magnitudes. Combined with protein assay technology, nanotechnology thus is promising to bring more sensitivity into in-vitro diagnostics and measurements. It might be relevant here to elaborate on the essentials of bio-applicable QDs in order to get the underlying flavor of this development.

QDs are nanocrystals of semiconducting materials having typical size in the range of 2 to 10 nanometers in diameter. The bio-applicable QDs are essentially inorganic fluorophores having much improved optical properties than the conventional molecular fluorophores, e.g. fluorescein. They have a core-shell structure where the core is made of one semiconductor material which is covered like a shell by another semiconductor having a comparatively larger band gap (characteristic property of a semiconductor) than that of the core. It is almost universal that all QDs can be excited by use of UV light. The shell material is chosen critically to enhance the yield and stability. Normally, these particles do not dissolve in water, but in order to make them bio-compatible, silanization and polymer-coating is practiced. Also, they conjugate with many different types of bio-molecules, e.g. oligonucleotides, antibodies etc. When they absorb light with typical photon energy larger than the band gap of the shell, an electron-hole (exciton) pair is produced. These excitons have longer life time (10-40 nsec) within a QD because of special nanoscale size as compared to their life span in any conventional molecular bio-level (<10 nsec). When excitons de-excite and the electron-hole pair recombines, a photon is emitted that produces strong fluorescence signal having narrow and symmetric band. The emitted light is what characterizes a QD and its characteristics depend on the physical size of the particle. The emitted light also carries information about the environment that a QD resides at that particular time. If the physical size of a QD is comparable to the typical distance of an electron-hole pair, the emitted photon energy is quantized and this phenomenon is known as *quantum-confinement*. Thereby, the energy of the emitted photon can be related directly to the size of a QD. It hence establishes a direct control of desired effect by a *nanotechnology variable*, as discussed previously to explain Fig. 6. The light signal emitted by such a QD can be measured employing various optical instruments, e.g. different types of spectroscopic methods or fluorometers, as per necessity.

This novel feature of QDs has been very effectively utilized to develop various in-vitro diagnostics. For example, using relevant antibodies conjugated to QDs, simultaneous detection of Ricin, Choleratoxin, Shiga-like toxin 1 and Staphylococcal enterotoxin B have been made possible, in multiple immunoassay. Quantities as low as 10 nanogram (1 nanogram is one billionth of a gram) per milliliter cholera toxin could be detected. Antibodies against Hepatitis B and C viruses could be detected with a picomole sensitivity using antigen coated QDs embedded in polystyrene microbeads in a multiplex diagnostic system. Moreover, in genomic analysis and nucleic acid detection, significant results are obtained using QDs. These examples demonstrate the utilization of QDs in bio-labeling and the immense potential to apply nanotechnology for in-vitro diagnostics. The success of these methods including the special instruments developed to carry the studies will eventually create a substantial market in near future. Per se, existing medical diagnostic tools make a billion dollar market at present!

Metal nanoparticles (NP) is another nanotechnology application for in-vitro diagnostics. In general, nanoparticles are agglomerates consisting of countable number of atoms and have a size smaller than the typical bulk grain. In principal, nanoparticles of all metals in the periodic table can be synthesized. However, the techniques do vary for different metals and also for different size range of the same metal. Their typical size range would be 10 to 300 nanometers. Smaller particles of sub-nanometer to 10 nanometer sizes are typically referred as nanoclusters, and their size is normally measured by the total countable number of atoms or molecules that each one contains. In general, metal nanoparticles are highly reactive. At very small sizes, the properties of each of them could drastically vary from each other. This range is considered non-scalable. Little larger sizes though can be correlated in terms of specific individual properties. Even though small ones are distinct in their particular non-crystalline geometric structures, they might be approximated to be spherical in simple measurements of size distribution as per early theories. These particles have *surface-to-volume* ratios much higher compared to that of the bulk metal. In bulk, most atoms are within the bulk and a percentage only constructs the surface. On the contrary, in nanoparticles, most of the atoms are on the surface and a percentage only constitutes the bulk! As they grow larger, the number of atoms within the body increases finally to reach the bulk limit. Therefore, the nanotechnology variable *surface-to-volume ratio* appears to be a significant parameter to correlate many new and novel properties of NPs. Such as, the non-crystalline structure, aggressive reactivity, special optical and electrical properties, larger magnetic moments, superior catalytic coefficient, special thermal properties etc. Special shell-like configuration could also be obtained in specific size nanoparticles. For example hybrid gold NPs are synthesized with a thin shell of gold atoms surrounding a dielectric core of either an insulating material or silica.

Some novel properties of NPs may be utilized for in-vitro diagnostics. For example, surface *plasmon* resonance, larger magnetic moment etc. In fact, plasmon-plasmon interaction is useful for colorimetric detection and enhanced magnetic properties could be utilized for advanced MRI in difficult situations. However, the main trick of the trade in biological applications is to attach or bond the NPs with the bio entities, e.g. DNA, protein, enzymes, antigens and antibodies etc. Gold NPs have been demonstrated successfully for various in-vitro detections. When electromagnetic radiation of wavelength larger than that of the physical size of gold NPs falls on it, the free electrons across the NP undergo coherent resonant oscillations known as Surface-Plasmon Resonance (SPR). This results in strong optical absorption, which also depends on the environment of the NPs. As external bio-molecules are adsorbed on the NPs, plasmon-plasmon interaction shifts the absorption band causing a visible shift in color. This property could be used for colorimetric detection of analytes by measuring the shift in the values of the refractive index of the environment that the NP is situated in. For high sensitivity detection, gold NPs are found useful in labeling DNA and certain proteins. The labeling could be utilized for immunoassay, molecular diagnostics and imaging. It had been possible to develop a microchannel immunoassay to detect E. coli and H. pylori antigens with 10 nanogram sensitivity. This type of detection methodology is very promising for miniaturized device applications. Assay has also been developed using protein chip enabling simultaneous detection of Hepatitis B and C viruses. 1-200 picogram sensitivity detection of human chorionic gonadotropin and total prostrate specific antigen have been made possible by labeling gold NPs with secondary antibodies. Hirsch et al. demonstrated a whole blood analysis using gold NPs in an immunoassay capable of detecting rabbit IgG with sub-nanogram sensitivity in multiple media. Bio-barcode assay have been proposed with zeptomole DNA sensitivity that is highly promising as a future alternative of PCR technology. Protein detection with auto-mole sensitivity using BCA has been demonstrated as well. Such amazing achievements in bio-detection could be possible using particles that contain just a few molecules!

5.4.3. In-Vivo Imaging

As the name suggests, the technique is to see through live system or image an internal organ of a living system, without perturbing the normal modes of its functionality. The light emission properties of QDs and metal NPs are useful in this regard to probe or examine specific organs of a living system including the human body. QDs can also be used to track the journey of a drug molecule starting from the point of intake as it traverses through the digestive pathways till it gets absorbed within the tissue. The advantage of using QDs for imaging is that they remain resistant to bleaching for hours, thus allowing an acquisition of well contrasted images. Important observations like, whole cell labeling for pathogen detection, cell tracking and cell lineage studies etc. need longer imaging time. In fact, only a small quantity of sample is required in immunofluorescence studies. It has been reported by several workers that even a single QD could be observed at immunocytological conditions which provides an extreme sensitivity by one QD per target molecule. Additional precautions are mandatory especially because the procedure shouldn't harm the living organism itself. Unfortunately, this is still a major challenge as in general, QDs are found to be highly toxic! Further, they need to be washed out of the system once the image scanning is over. As a matter of fact, the mechanisms for such actions are not yet understood completely as to be considered foolproof. Cytotoxicity measurements have shown that if the core/shell material in a QD is protected, the interference with cell viability is lesser. New designs with total biopassivation might evolve which will prevent the active material to come in direct contact with the cells or fluids so long it is within the living system.

Gold NPs on the contrary have been shown by many researchers to be non-cyto-toxic. These entities are therefore more appropriate for in-vivo imaging applications. 12 nanometer diameter gold NPs have been successfully used to image and measure the red shift for cultured cervical epithelial cancer cells. An example of another tricky application is found in the treatment of superficial growths, like non-malignant tumors. Non-surgical and less painful localized destruction could be achieved utilizing the light absorption property of gold NPs. Microinjected gold NPs could be excited by short laser light pulses to generate excessive heat causing destruction of the tumor cells very locally, because these cells are found to be more heat sensitive. Such procedures are referred as Photo Thermal Therapy (PTT). Another class of NPs used for imaging for quite a while now is the super-paramagnetic ones. They exhibit large magnetic moment under the influence of a magnetic field as compared to normal paramagnets. They are advantageous in exhibiting large contrast in MRI imaging and could be more widespread within the tissue because of their smaller size. Iron oxide NPs are used for this purpose however by stabilizing them within polymer matrices, e.g. polyethylene glycol, starch or siloxane etc. FDA approval (USA) has been issued in this regard to use the commercial variety of iron-oxide NPs available as *Feridex* to observe the MRI of spleen, liver, gastrointestinal tract etc. Its utility has been demonstrated recently even in cardiovascular imaging. Glucose capped gold NPs are also found to have improved cellular targeting and radiation sensitivity having clinical implication for radiation-resistant cancer cells like human prostate cells. Most of the clinical applications of QDs are listed in Table II as a summary.

Finally, the question is what we gain at the end by implementing such sophisticated techniques like in-vitro diagnostics and in-vivo imaging? Actually, medical diagnostic procedures together with tools and kits, reagents, maintenance and technical supervision make more than a billion dollar industry around the world. Speculations are that the nano-technology market is expected to increase by 30% per year. Already some QDs and gold NPs are being made available at open market and are being used in detection of HIV, ovulation and pregnancy tests. Though sophisticated, these methods have very high sensitivity, small turn-around time and are cost effective because it gives more precise results. Moreover, the cost and the hassles of performing several tests could be reduced through one single test. Further, because of extreme high sensitivity, it is useful for pre-symptomatic investigations that determine preventive medications or procedures. Therefore, these methods offer astounding promise to the field of clinical diagnostics. Nevertheless, completely standardized procedures are yet to be consolidated and such applications also have to overcome various societal and safety issues as well.

Table II

Quantum Dots

	In Vitro Imaging	In Vivo Imaging	Bio-analytical Assays	Bio-sensors	Fixed Cell Imaging
Method →	Fluorescence detection	Fluorescence, MRI and PET detection	Fluorescence detection	Fluorescence detection	Fluorescence and TEM detection
Applied for →	Live cells	Live Tissues, Organs and Tumors			Cells and Tissue sections
Clinical application →	Membrane surface	Photo-induced therapy	FC analysis	QD-tagged anti-bodies	Immuno-Chemistry
	Intercellular organelle	Aiding optical surgery	Optical coding of cells	FRET-QD Sensors	Multiplex FI
		Therapeutic agent	Micro-arrays	QD-encoded micro-beads	

Source: X. Michalet et al, *Science*, vol. 307 (2005) p. 538.

5.4.4. Regenerative Medicine, Xenotransplants

The field of Regenerative Medicine is actually the modern version of Culture of Organs that has its root way back in 1938. Eminent French surgeon and biologist Alexis Carrel, who won the Nobel prize, produced the blood vessel graphs for human body and devised technology that is used even today for suturing blood vessels. The field is at a very advanced stage and currently, artificial or cultured organs are produced routinely for replacement of the damaged ones. The research activity in this field is very important to substantiate the shortage of natural organs with the artificial alternatives. Dr. Anthony Atala, Director of Wake Forest Institute of Regenerative Medicine, Winston-Salem, NC, USA, has very nicely summarized the issues and advances in this field in a TED talk. Many different areas are involved in regenerative medicine, e.g. smart bio-materials that are used as scaffolds, cells from patient's body or stem-cell population (or both) are all used in hybrid formations. As Dr. Atala said, there are many challenges in the field, e.g. the design of materials that requires innovative technology, growing cells outside the living system that needs extremely stringent conditions to be sustained for very long period, copying the blood vessel graph onto an artificially cultured organ and initiating the supply of blood into it. Smart biomaterials could be used to create the scaffolding for any particular organ which is then coated with specific types of cells cultured from a good tiny piece of original tissue under artificially controlled environment. The engineered organ is subsequently preconditioned under the body-like atmosphere before implanting into a patient's body as a replacement of the defective organ. Nanotechnology could be useful to contribute in designing the complicated blood vessel graph onto the biomaterial scaffolding to be coated with cultured tissues. This would become plausible as we gain expertise on molecular level engineering. A specific computer grafted design of any structure might be built in real scale molecule by molecule, thus creating the entire organ artificially! With the availability of complete genome data for interpretation and use, regenerative tissue development would become more precise. This would help to develop treatments that are patient specific. Recently, scaffolding has been materialized utilizing the rejected organs in which at least the blood vessels are not bad enough to be reused. This scaffolding is then re-perfused with the cultured or stem cells to create the complete organ. Remarkably, 3D ink-jet printing technology has also been introduced to perfuse the scaffolding with cells! Research work is in progress to create artificial organs using computer-controlled morphometric imaging analysis and layer-by-layer 3D reconstruction through jet-printing technology. This also gives another possibility for nanotechnology applications where nano-lithographic procedures might be thought to be implemented as an alternative. Together with smaller chip array, it should be possible to have faster analysis and fully compatible nano-engineered artificial organs for individual patients.

Developing artificial organs is nevertheless a costly affair, but it is specific and avoids many anticipated problems that medical science encounters in xenotransplants, practiced over quite a long time now. Actually speaking, xenotransplant is a kind of one way traffic! It seldom happens that any animal has been cured using a human organ, whereas an artificially synthesized organ could also be used to save the life of the pets. Therefore, tissue engineering, implants and its related devices or equipment has promising commercial importance and is the most viable option of investment, more in the developed world as its population grows older. Besides, any replacement using patient's own genetic material is superior because of compatibility issues and chances of rejection by patient's body are reduced substantially. Thereby the procedure has application in artificial skin, tissue reconstruction, on-spot wound healing by tissue filling etc., as has been demonstrated through prototype *scanner technology* by Dr. Atala and his group. The impact of nanotechnology will be in all the areas of regenerative medicine through advanced nanoscale devices capable of handling issues of molecular engineering, genetic engineering, immunological characteristics of diseased cells, repairs of bone and cartilage, vascular surgery and engineering of the nerve cells, etc. Canadian Institute of Health Research (CIHR) published a commendable work in this regard. Intelligent Nano, a company in collaboration with National Institute of Nanotechnology (NINT) in Edmonton, Alberta, has applied nanotechnology for regenerative medicine by devising *SonaCell* that would be used in simulation of stem-cell proliferation, fermentation of antibiotics and increasing bio-fuel production (March 17, 2011).

5.4.5. Biomimicry

It is the subject of *innovation by emulating nature* popularized by Janine M. Benyus in the book *Biomimicry: Innovation inspired by Nature*, published in 1997. It is to learn the hierarchical functionality of biological systems and emulate it onto artificial systems. In other words it is to mimic the functionality of existing biological systems and build things for our specific use.

With the advent of our understanding in nanoscale science, it is now possible to appreciate the fundamentals of inimitably intricate and highly ordered structural functionality inherent in biological systems. The knowledge of biological systems at molecular level has paved the way to appreciate (1) the immense order in self-assembly which is like instructions programmed within the molecules and (2) the hierarchy, not only in structure but also in the execution of command sequences. For example, two complimentary DNA strands would form a double helix or a linear amino acid strand would fall on itself to form a 3D protein structure. It is an innate character of functional specificity of the bonds, electronic structure and the implied forces therein. Complex structures are formed at nanometer scale through combinations of atoms and molecules, that at micrometer scale manifests as various types of cells responsible for higher order phenomenon, which then assemble to form further higher level through associative processes required in different somatic functions. All integrated is what creates a biological life! The basic functional level of this multifold is at nanometer scale where independent functions could be back tracked and correlated as instruction codes for the next higher level. Salvatore Santoli, in his scholarly article entitled *"Why nanostructuring and nonlinear informational dynamics for biomimicry?"* speculates that cells apparently work as connections between the macroscopic world and the microscopic nanoscale sub-cellular entities. The task for biomimicry is in fact more towards application by way of answering the question, "Can we exploit the special information from bio-systems in designing a functional molecular device artificially that can work through our command structure?" In one sense, this is more pragmatic an approach towards realizing Feynman's vision of making the ultimate molecular machines—the objective that initiated nanotechnology! Drexler's molecular engineered machines in the virtual world of computers might be too distant to materialize in reality, but machines out of molecules could perhaps be realized by engineering, e.g. DNA or protein at molecular level. Therefore, imitating nature under the folder of *biomimicry* has immense prospects. It furnishes and fathoms an alternative to the direct bottom-up approach of molecular assembly in building nanoscale machinery. Here we imitate to procreate similar functionality! However, the theoretical understanding of the issues in biomimicking is rather intricate and involved because of the fact that structure→compactness→function in biosystems are inseparable from each other and the dynamics is highly nonlinear. There exists vast network of molecular interactions that are open and wide apart from thermodynamic equilibrium. There is continuous exchange of information that involves exchange of matter and energy with the surrounding environment. The character of biological information is in true sense unlike the code language for computer programming, which does not have any sort of involvement with the matter it is meant to work for. For example, a software routine executing the charging-discharging of a battery cell has nothing to do with the battery material. Though the function is automated, it is not intelligent to learn by itself what to do if something suddenly goes wrong! The information within biosystems, on the contrary, is generated newly every time an interaction happens and there is no copy retained! It considerably simulates the environment and builds its structure due to the coded interactions. The intelligence level is far superior to what we have as our computer program. Further, there is innumerable parallel processing that even the same interaction might result into different responses at different environments! This evolutionary behavior of the level of complexity underlying the intelligence of biosystems is a highly non-linear phenomenon and is self-organized at a far-from-equilibrium situation. Moreover, this behavior is organized in a hierarchical structure from molecular scale creating larger macro scale structures which again does not remain just like a sum of individual parts. Actually, the whole is more than the sum of its parts in biological systems as depicted rightly by Raiesdana et al. (see Bibliography). Tackling such complexity necessitates the study of higher degree non-linear dynamics knowing the control of chaos mathematically, in order to design and model specific functionality. Some examples of biomimicry will be discussed in the following to highlight the inimitable internal connections with the material world at nanoscale.

5.4.5.1. DENDRIMERS

One approach to biomimicking experiment is with synthetic polymer molecules that are found suitable to mimic certain molecular functions. Research on dendrimers is important in that. In general, there are three major macromolecular architectures of synthetic polymers: Class I: Linear, Class II: Cross-linked or Bridged, Class III: Branched. Architecture of dendrite is grouped in Class III. The topology of this architecture is very common in biological systems in almost all length scales, e.g. tree branching, roots branching, human anatomy, various internal organs, i.e. lungs, livers etc., neurons and nerve cell structures, everywhere! The pervasiveness of this topology leads to speculate that such architecture might provide maximum interface for any distribution network and it has a strong rapport with the biological hierarchy. Polyamidoamine (PAMAM) is perhaps the first complete dendrimer family to be synthesized, characterized and made commercially available. They have extraordinary structural control at nanoscale and they mimic the structures of globular proteins.

Because of outstanding dimensional control and electrophoretic properties, PAMAM is referred as *artificial protein*. For example, Insulin, Cytochrome C and Hemoglobin are approximately the same size respectively as Ammonia core 3, 4, and 5 generations PAMAM. Lipid bilayer membranes are equivalent to 5 and 6 generations of this dendrimers. DNA duplexes form highly stable complexes with 7 to 10 generations PAMAM that has application possibility in the delivery, storing and condensing of specific drugs. The comparative physical characteristics have also given commercial use of these dendrimers as replacements of globular proteins in immunodiagnostics and in-vitro gene expression applications. The advantage with dendrimers is that their three dimensional structures are robust mainly because of very strong covalent bonding. This electronic characteristic actually makes them fundamentally different from globular proteins which have tertiary structures resulting out of complex folding of linear structural sequences, making the proteins fragile and highly dependent on denaturing conditions. Therefore, dendrimers cannot actually have the biological activity that proteins are responsible for within a cell though the 3D structures of dendrimers contain both interior hollowness and well defined surface functionality like that of proteins. Because of the robustness they might be good candidates for creating nanoscale scaffolding required for regenerative medicine as discussed earlier. Hence, these polymers have promising role in the development of new biomedical applications either by fabricating nanoscale devices or devising strategies for treatments. Researchers have shown the unique drug loading capabilities and also the targeting features of dendrimers. In addition to drug delivery, their usefulness in gene-transfixion, pathogen-pacification, immunodiagnostics and as contrast agents in magnetic resonance is highlighted as well. Nevertheless, dendrites are fundamentally different compared to proteins which are made of amino acids joined covalently through peptide bonds. In larger proteins, many such smaller units could remain bonded non-covalently between themselves. Dendritic polymers actually are passive entities and far off to be considered biologically active.

5.4.5.2. VELCRO

Various other forms of biomimicry are being investigated and tried out by researchers because of strong speculations to achieve wonders. In fact, the most widely popular example of biomimicry in everyday use is *Velcro brand fastener* invented long back in 1941 in Switzerland by George de Mestral out of his canine fondness. Velcro is very widely used commercially in almost everything that requires to be fastened quickly. It is actually a hook and loop fastening mechanism in small scale. The idea occurred to him by investigating the seeds or burdocks that hooks on clothes and furs. The genius in him is to learn the technique from nature and reproduce in right scale with useful material viz. the polymer—nylon. It has enough strength for everyday common strapping purpose that a 2 sq. inch Velcro strip would support almost 79 kg. The most widely rampant use of Velcro today actually brings partial credit to Mr. David Letterman to some extent and his NBC late night show on Velcro jumping. It is an example of biomimicry where *nature's* ingenious way to spread seeds by hooking to close loops, has been emulated for our plentiful use.

5.4.5.3. GECKO TAPE

Imagine you could act like Spiderman to save someone from a bad situation! The solution might be round the corner in near future! Gecko tape, another example of biomimicry, has opened up a possibility of such an imagination to become a reality. This material is another example of inheriting the *nature's* intricacies in the form of an invention in which, the tiny lizard gecko's capability to walk across ceilings and up the walls has been imitated. The tape contains tiny nanoscopic hairs that mimic the hairs found in a gecko's feet. Millions of synthetic setae cover one centimeter area of the tape. These are made of synthetic kapton material by nano-lithographic technique and measures 2 micrometer in length and diameter, just like a gecko's keratin hairs. They are flexible and exert van der Waals force to provide a powerful adhesive effect. It is interesting to note that though van der Waals force is a weaker interaction, multiplied million times individually makes it great in effect. Because of the fact that we are now capable of making things at nanoscale size, it has become possible to make millions of them precisely within a centimeter. If 10^6 hairs are grafted in 10 milimeter space, each hair at least remains 8 nanometer widely spaced from each other! These tapes are most useful for underwater and space station applications.

5.4.5.4. ARTIFICIAL PHOTOSYNTHESIS

This is a very impressive and rapidly progressing field of activity in biomimicry. The knowledge of photosynthesis is replicated here for our benefits. In the beginning of May 2011, sensational news broke out around the world under the heading *Artificial Leaf* from Massachusetts Institute of Technology, Boston. It generated a feeling that pretty soon, nanotechnology would be able to synthesize food fundamentally as plants do, using sunlight and water; all worry on food crisis around the world will soon be a history; never more to be anxious even if there is no harvest because we would soon have the synthetic carbohydrates and proteins on board!

The origin of such a sensation has been an imaginative extension of a very important research that the then Professor of MIT, Daniel Nocera (currently, Prof. at Chemistry and Chemical Biology, Harvard University) published from MIT, which has been continuing dedicatedly for a long period. The findings were from the study of photo-electrolysis which means conducting electrolysis using light or photons, the particles of light. *Lysis* means breaking apart and therefore electrolysis stands for breaking something apart using electricity. In practice, it essentially means breaking water into hydrogen and oxygen. You simply need to pass electricity between two electrodes placed in the water (sparingly salted) to make electrolysis happen. The principle of electrolysis was first formulated by Michael Faraday in 1820. The question today is, "Can we achieve it by using light, preferably sunlight?" Why? Because this is what plants do in the first stage of photosynthesis within the green leaves. Then in the second stage, the production of glucose takes place by using CO_2 acquired from atmosphere. Is it possible for us to produce a mechanism to follow suit that would use atmospheric CO_2? This would enable us to at least mop-up some excess CO_2 thereby remedying the carbon emission problem even if it does not enable us to synthesize carbohydrates!

The first stage of photosynthesis in plant is a light dependent reaction where water is broken into hydrogen and oxygen using sunlight. This is *nature's* wonder that plants actually harvest the light energy. We are not yet capable of harvesting energy. This is where the importance lies in mimicking photosynthesis. Plants are capable of doing it using the catalyst, chlorophyll and other photosynthetic pigments and enzyme complexes. In fact, all the process happens within chloroplasts which are membrane bounded intercellular organelles with variable shape and few micrometer in diameter, like mitochondria. In this light driven oxidation-reduction reaction, two important things happen. First, H_2O donates electrons and second, they are used to make ATP, Adenosine Tri Phosphate. This is essentially a light driven phosphorylation reaction. Light energy is harvested in the form of ATP or NADPH and simultaneously O_2 is evolved in the process. These coenzymes work in cyclic manner like catalysts.

It has been tried to mimic this process in artificial photosynthesis. Blue dimer has been synthesized and dye sensitized cells are used but found not to be very efficient. Paul Kögerler used 4 ruthenium atoms. The genius of Daniel Nocera is to use a recently developed cobalt-based water splitting catalyst cobalt phosphate (Co-Pi) integrated into a silicon solar cell that resulted into water splitting and release of O_2 at neutral pH. This cell is what is referred as the *artificial leaf*. Oxygen bubbles are observed when these cells are submerged into water and exposed to sunlight! Here, the technology of silicon solar cell has been integrated to that of the synthetic (cobalt-phosphate) catalysis. The process creates an aggressive chemical environment by the release of oxygen, in that the action it has been made for, could possibly destroy the device itself (Frankenstein act). In fact, two teams could achieve water oxidation using silicon independently which most people did not attempt because it is known to form silicon dioxide. The credit for Prof. Nocera is that he developed this Co-Pi catalyst 3 years prior to conducting this research, which led him to succeed. The synthesis is however not simple in which, a layer of pure cobalt metal is first evaporated onto the cell electrode and then exposed to phosphate buffer solution under an electrical charge, to transform it into the Co-Pi catalyst. The surface becomes passive due to the formation of bonded layer of Co-Pi. Even then, this device is inexpensive because of the use of earth abundant low cost materials as compared to Dye sensitive cells or noble metal oxides or ruthenium atoms application. The ultimate aim of this biomimicry application has been to produce the full scale artificial leaf that can work as a reliable alternate source of power.

The second stage of plant photosynthesis is the light independent reaction or carbon assimilation (or fixation) reaction, which is sometimes misleadingly referred as the dark reaction. This stage is in fact driven by the products of light-reaction. ATP and NADPH are used to reduce CO_2 to form Triose-Phosphate, starch and sucrose with other products derived from them. The detailed biochemistry of this process is known and given in many publications and books. In brief, there are three stages in CO_2 assimilation process. The first stage is called the carbon-fixation reaction in which two molecules of 3-phosphoglycerate is formed from condensation of CO_2 with ribulose 1,5-biphosphate, an acceptor with five carbon atoms. In the second stage, 3-phosphoglycerate is reduced to six molecules of triose-phosphates. Five of the six molecules of triose-phosphate are used to regenerate three molecules of ribulose 1,5-biphosphate, the starting material and the sixth molecule of triose-phosphate is used to make sucrose for transport, hexose for fuel and starch for storage in the third stage of the process. This is how plants utilize CO_2 from atmosphere. The important thing to notice here is that the overall process is cyclical where continuous reutilization of biochemical ingredients and conversion of CO_2 happens like a superior process of churning out and production of multiple products. This happens incognito consecutive to the light dependent reaction that generates the proton. This is the uniqueness of this molecular scale factory! The cyclical nature is something that might be essential in mimicking for sustainable results. Nevertheless, artificial photosynthesis has not yet reached such sophistication. The main aspiration at the moment is to obtain an NADPH-inspired catalyst that should be capable of recreating the natural cyclical process. Brookhaven chemists have been trying with a ruthenium-based complex which behaves somewhat similarly giving the proton and two electrons that can convert acetone to isopropanol. Current research is targeted towards obtaining light generated proton and then to produce a fuel from CO_2. Therefore the industrial objective of artificial photosynthesis project is twofold; (1) to obtain light generated proton using special catalyst-coated solar cells and convert to molecular hydrogen to store in hydride cells as fuel for next generation transport or automobile industry; (2) to perform catalyst dependent conversion reaction to utilize CO_2 and produce bio-fuels (like bio-diesel) that helps also to reduce the CO_2 content in atmosphere at the same time. Sustainable hydrogen production in large scale from water just utilizing sunlight is of key importance for clean and affordable alternate energy in future. Also, both of these efforts would eventually mitigate carbon emission and global warming.

5.4.5.5. LOTUS HYDROPHOBIA

A lotus leaf cannot be made wet! They are super hydrophobic. There are also a variety of plant leaves which exhibit similar nature. Water dropped on them beads and roles down even carrying the dirt out of it. Whatever might be the reason for this super hydrophobic nature in plant world, researchers are trying to mimic this extraordinary property for metals and plastics! Advantages are many. For example, imagine that your next vehicle has such a windshield and such a metal body! Neither do you need a wiper periodically blocking your vision while driving in heavy rain nor do you require waxing of the body of the car to prevent your chosen color getting worn off. There will be many other commercial applications for such water repellent materials also. For example, in de-icing of aircrafts which is sometimes a threatening problem in high altitude and high speed flying, especially for fighter jets; in protecting metals used in contaminated environment where plenty of moisture gathers for example, blades of steam (or gas) turbines; in protecting metals used in salty sea water, e.g. shielding the body of ships from corrosion etc. GE researchers are on their way for this type of biomimicking work.

Lotus leaves are coated with very tiny wax crystals about tens of nanometer in size that hold the water as perfect spherical beads. Following this, two approaches have been devised. One, texturing of a metal surface and then put a water-repelling chemical coating on top or two, texturing the coating itself and leaving the metal untouched. However, there might be no universal solution for all metals. Factors like the robustness of the material to be coated and the sticking coefficient of the coating etc. would be the primary deciding factors in performance. One procedure might be superior over the other depending upon the material to be used. However, we still have to wait before we can save the dollars spent in the car wash because such materials are yet to be realized.

5.4.5.6. SELF-HEALING PLASTICS

In today's world plastics are part and parcel of everyday life. The main reason of popularity of plastic is that it is insulator, lightweight and cheaper than metal; does not make cranky sound on impact and can be made colorful as well. It is breakable but sturdy enough for common use. But imagine a time, when you have some food items in a plastic container and you are out for a picnic; this box accidentally slips off the car; there are some cracks but by the time you are done for the day, the cracks are healed by themselves! Well, these are not the plastic that we have our experience about. These are new polymer composites that mimic our blood coagulation property. We know that there are particles called platelets in our blood stream. Whenever there is any cut or bruise, platelets present in blood starts functioning; they condense and creates a block to prevent further loss of blood from the body. This self-healing property inherent in biological world is being mimicked in material world. Composite materials are being synthesized that contains tiny hollow fibers filled with resin which is released under stress creating a scab by itself on any crack in the material. The crack is healed as if with a spontaneous self-generated protection as strong as the material itself. Such bleeding plastic with built-in damage fixing property is not really imaginary any longer but are expected to be products in the coming days! Yesterday's science fiction as today's reality! Imagination seen in today's animated films like extraterrestrials being torn apart by massive bullets while their body material flows in to unite again and reconstruct the body—maybe such a scene won't create a shock anymore, but be accepted as tomorrow's reality. It's merely a matter of time!

There is immense utility of such a material depending on its strength and how it compares to metal. For example, in making lighter aircraft and for self-healing tiny cracks caused through rundown wear and tear in an aircraft made of such a material, because the micro level inspection won't be necessary after each exhaustive long run where the thermal stress takes serious toll on the material over the years. Bringing this technology into the aviation industry would create millions of dollars in savings by having a reliable safety feature. In addition, new composite material in which a network of nano scale fiberglass tubing is completely embedded has been proposed. Such tubing is supposed to be filled in with the crack healing substance so that a complete network, similar to the network of blood vessels in our body, could be built. Such materials will be inactively alive so that refilling and reuse is amenable and any single structure can repeatedly heal throughout its lifetime. The endeavor thereby explores the potential to develop biological-type functions in man-made materials!

5.4.5.7. BIOSILIFICATION

The name biosilification suggests some material synthesizing process. In fact, research in this subject is to develop the fabrication technology through the study of diatoms and in some cases, by using them as well. Diatoms are photosynthetic algae, unicellular and a most common type, phytoplankton, available in almost all aquatic and moist environments. The fossil evidence shows that these species exists since the early Jurassic period. There are more than 200 genera of diatoms identified so far. They are non-motile and live forming colonies in ocean, fresh water, soil, dump surface or anywhere moist. They have a robust and unique cell wall made of hydrated silicon dioxide (silica). This siliceous skeleton is a fundamental characteristic of diatoms and is termed as *frustules* by biologists. Actually, both benthic and planktic forms of diatoms are found with a typical size of 20 to 200 microns. Sometimes, even 2 millimeter long sizes are found. These algae are producers in food chain and form endoplasmic cysts storing oil rather than starch. It secretes silica at some stage of the life cycle through a bipartite cell wall. Primarily it has asexual reproduction through binary fission in which the daughter cell receives one of the 2 frustules of the parent cells. However, sexual reproduction and auxospore formation is also observed in some cases. Auxospore is a temporary spherical organic membrane that protects the initial stage of growth in sexual reproduction.

Diatoms are being studied since late eighteenth century. In fact, it used to be the most popular sample for people in early nineteenth century involved in microscope building in order to determine the resolution of the equipment. People even produced several hand-drawn illustrations of diatoms in monographs just by looking at them under the microscopes in a bid to improve the resolution of the microscope using better optical components. Nevertheless, there is a renewed interest today from a different perspective and diatoms are becoming more and more important for nanotechnology!

When diatoms die, their soft internal parts deteriorate, leaving the hard porous skeletons or frustules behind. In fact, huge number of such skeletons forms thick deposits on the floors of coastal habitat or in former aquatic inland. These dead remains are used commercially in a number of ways, like fine abrasive such as silver polish and toothpaste, as filters, mineral fillers, insulating materials and anti-cracking agents etc. These are coarse applications in which the intricate nanoscale internal structures in frustules are of no importance. However, with the advent of our technological capability, we can now appreciate these intricate designs by bringing them into more sophisticated applications, especially to fabricate nanostructures where frustules are used as templates. Complex, spatially patterned micro- to nano-sized structures can be constructed by means of thermal deposition of metal. Particular interest has been given to form gold nanostructures. Highly regular three dimensional silica nanostructures having nanometer dimension of frustules therefore opens up ample opportunity for biomimetic fabrication of materials. Why is this important? There are a number of top down nanostructure synthesis technique like, electron beam lithography, laser ablation, electrochemical etching, template synthesis, etc. which are expensive in terms of investment and running cost and they are also time consuming processes. Use of nanoscale structured templates for synthesis has emerged popular because it is less expensive, versatile and highly reproducible with a potential to produce ordered nanostructures in mass scale as well. Porous membranes, zeolite and crystalline colloidal arrays have been used before for such fabrications. Templates obtained from different biological species, e.g. sea urchin, pollen grains and cellulose fibers etc. have also been used. But the intricate silicate-based architecture available in many different three dimensional sizes and shapes in diatom frustules offers the largest opportunity of having extraordinary templates to create highly ordered nanostructures. Diatoms can be routinely grown under laboratory conditions and unique morphologies can be obtained thereof. The use of diatom frustules as templates therefore offers a highly potential alternative to expensive lithographic techniques. Fabrication of gold nanostructures has been very successful using this methodology. Gold atoms are thermally evaporated on the diatom frustules, followed by a stripping process that results in the fabrication of a film. An inverse gold nanostructure, like a transcript is created on deposition of gold atoms onto the pores of the frustules. Such gold nanostructures have been fabricated by Dusan Losic et. al. (see bibliography). It is important to take proper precaution in preserving the nano-topography of the internal structure within frustules in order to create a highly ordered inverse replica. Two types of formations can be created. One that replicates the planner surface structure and another, which copies any particular three dimensional curvature. The latter is important in view of the fact that diatoms have many varieties of valves that is used to regulate the secretion of silica while they are alive. The structural remains of these valves in frustules gives opportunity to construct highly ordered miniscule valves using proper synthetic or biomaterial. Apart from that, a number of potential applications of diatom frustules have been proposed by several researchers e.g. in optics, electronics, bio-photonics, catalysis, lubrication and drug delivery etc. It has been observed that diatoms at certain stage of its lifetime secret amorphous silica. This is a great resource for biological production of amorphous silica. Silicon is the second most abundant element on earth and is one of the most widely used semiconductors that is as well used in making glasses, ceramics, plastics etc. and many other products. Diatoms can be easily grown and used in chemostat cultures for mass scale biological production of amorphous silica. These biologically produced silica exhibit a genetically controlled precision in nanoscale architecture that our existing engineering capabilities are barely capable of achieving! In addition, the production can very well be in large quantities by actually investing much lower amount of capital unlike any other chemical or engineering synthesis techniques. Tons of silica deposits can be obtained on controlled water reserves, protected natural lakes and on the floor of the oceans.

Nevertheless, what actually is the far-fetched implication of getting involved in such activity? Silicon biotechnology has just started to evolve. Genetic level study of single cellular diatom would essentially reveal how genes and molecular mechanism controls the biological nanofabrication of silicon-based materials. Such discoveries in fact interface between the biological and the inorganic world by establishing a link between siloxane polycondensation and biopolymers. It would lead to understand the role of enzymes as dynamic catalysts in the process and achieve biotechnological control that could hardly be imagined before!

Diatoms also have possible application as photonic crystals. The amorphous silica frustules are consisted of two halves, called the *thecae,* which can be divided into a valve and a grid band. In some cases, there may be more than one grid band. The exceptional periodic pattern of holes in a slab, both valve and grid bands can be regarded as slab waveguide photonic crystals with distinct symmetries and spectral ranges. It has been shown that the optical properties of diatom shell can influence incoming light by coupling into waveguides with distinct photonic crystal modes. Therefore, diatom shells can have application as photonic crystal slabs. The advantage is, it is extremely easy and cheap to reproduce compared to any sophisticated mechanical fabrication of such crystals.

Another intriguing recent discovery about diatoms in connection with nanotechnology application must be mentioned in credit to Georgia Tech researchers who discovered that diatoms also store blobs of dense concentrations of phosphorous as polyphosphates. Diatoms transport phosphorous in the form of intercellular polyphosphates as they sink from top to the bottom of the ocean. Analysis showed that some parts of polyphosphate blobs have the structure of the mineral known as apatite and some other parts are like a transition material between a normal and the mineral. This discovery is important to tally the phosphorous account in ocean. It has been a mystery to scientist and oceanographers that the amount of phosphorous present in sea is always much less than that is washed off from rivers! This discovery might lead to an answer to this question besides knowing on the intricate mechanism of chemical transition in phosphorous mineralization in *nature*. It might be worth investing now in a culture of specific variety of diatoms in order to recycle back and reproduce the phosphorous being washed off on a daily basis around the world.

There are many more examples of biomimicry research and applications and the field as a whole is to be counted as highly interdisciplinary. For example, whale-power wind turbine, bionic car design, morphing aircraft wings, friction reduced swimsuit like shark suit, glow-fish, insect inspired autonomous robots, butterfly inspired displays, naturally cooled building designs mimicking large termite towers etc. It is not that all are connected to nanotechnology at the moment or to be married with nanotechnology very quickly. However, the knowledge of the science at nanometer scale and our advent in the capability of manipulating atoms and molecules, would certainly lead to further achievements in biomimicry applications more profoundly and rapidly.

5.4.6. Robotic Surgery

The very word *robot* is actually incorporated from a 1921 Czech play *R.U.R.* by Karel Capek and means "forced labor." It has subsequently evolved to mean dumb machines that perform menial, repetitive tasks. Today, robotic technology is much more improved though robots are still unintelligent machines but utilized to perform highly specific, precise and even dangerous tasks, e.g. manufacturing microprocessors, exploring deep sea and working within hazardous chemical environment, just to mention a few. Application of robots in medicine is comparatively a recent phenomenon that dates back to 1987 when the first laparoscopic cholecystectomy has been performed. In fact, this fact marked of a new era in medical practice, especially in surgery that goes by name *minimally invasive surgery*. There are many advantages compared to that of the conventional methods that even the medical insurance companies are instrumental in urging physicians to recommend it more to the patients! The benefits could be listed as, (1) smaller incision that reduces the chance of infection and earns savings in terms of utility and time; (2) shorter hospital stay which might even be avoided at times; (3) significant reduction of after surgery convalescence; and (4) overall, the failure rate in common surgery cases are highly reduced. All these factors combined together have effectively reduced the cost in common surgeries. Over the time, laparoscopic methods are extended to other surgeries in which the procedure could be performed with minimal incision or through the natural piercing in body. Other than removing gallstones, it has also been applied for prostate surgery, ovarian cyst, fallopian tube removal, etc.

However, surgeons encountered some limitations in laparoscopic methods:

1. The method is somewhat counterintuitive, i.e. the instrument has to be moved in the opposite direction to reach the target than what is shown as the course of action in the two dimensional video monitor. The hand-eye coordination is compromised in moving form a direction opposite to the target. This is unusual to the sensory perception of human brain and becomes taxing in long and continuous surgical procedures. Such situations are therefore risky in life-saving cases or in treating very critical patients.

2. The instruments have typically four degrees of motion compared to seven degrees of movement in our wrist. This also imposes a restriction that demands good amount of practice to become versatile. Any occasional physical tremor in surgeon's hands is easily transmitted through the length of the rigid instrument. Since a surgeon has to use the machine while in standing position, the chances of small amount of shake and vibration cannot be ruled out.

These drawbacks actually limited laparoscopic applications to be considered for delicate dissections and for critically conditioned patients even to perform a simple cholecystectomy. Thereby, the necessity is felt to extend the scope and capability of laparoscopic machines making them more efficient.

Initiation of robotics in surgery can actually be identified to have begun as an improved measure in laparoscopic methods. The overall benefits that laparoscopic techniques provide in saving time and money has been the compelling force for newer innovations making it more user friendly, safe and versatile. The use of robotics in surgery has actually begun in 1985 with the introduction of PUMA 560 that has been used to perform precision neurological biopsy. Then it was applied for transurethral resection of the prostrate in 1988. However, within a few years PRO-BOT was developed particularly to perform transurethral resection of the prostrate. Almost at the same period, the Integrated Surgical Supplies Ltd. in California came out with ROBODOC, the efficient robot doctor to perform precision hip replacement surgery. This machine is very successful and received the approval from the US Food and Drug Administration (FDA). On the other hand, NASA developed in parallel the technology of virtual reality to be applied for telepresence surgery, initially over a room or over a few rooms apart. It generated a platform where the surgeon could have the feel as if conducting the procedure directly on patient's body. Surgeons and endoscopists quickly realized the tremendous potential in these machines to perform flawless rapid and precise procedures and thus to get rid of all the drawbacks encountered in conventional laparoscopy. The armed forces department quickly realized the potential application for battlefield and in avoiding the death during transport from the base camp. Wartime mortality could be reduced by taking preventive measures against the exsanguination of wounded persons before reaching the hospital. The course of further development followed through the AESOP robotic endoscopic system that was voice enabled to receive commands from the surgeon to manipulate the arms and the endoscopic camera.

The most sophisticated instruments in this class today are the *da Vinci* surgical system developed by Intuitive Surgical and the *Zeus* system from Computer Motion. The parallel development in two directions described above has been merged together in these systems and designed to facilitate complex surgery using minimally invasive approach. In honor of Leonardo da Vinci's invention of the first robot and his extraordinary recreation of precise human anatomy in three dimensions, Intuitive Surgical named their system as *da Vinci*. It has four robotic arms interacting with the surgeon sitting in a control console. Three arms are the surgery tools and the fourth holds the endoscopic camera that gives full stereoscopic vision from the console. These robotic hands have all seven degrees of freedom in movement like human wrist. The surgeon has to sit in the console and look through the eye holes into the full three dimensional image while maneuvering the three arms of the robot using two foot pedals and two hand controls, which is certainly no less challenging than driving a high speed car on a busy highway or even piloting a fighter jet within an enemy area! Dexterity is extremely important. Surgeons might have thought that a robot would reduce their manual function, but now it is not only a job of two hands, but two legs as well! Precision is not easy to achieve just by sitting. The extraordinary ergonomic design of the system allows the surgeon to control everything by himself including the positioning of the camera while his eyes and hands are positioned in-line with the instrument. Surgeon has to simply move his hand in order to reposition anything like how we are used to doing normally. However, the design is such that at no time can a surgical robot become autonomous. The entire functionality works as a master-slave relation in which the robot performs only as a slave. This unit has been successful in prostatectomies, gynecological surgeries and cardiac valve repair. The superior visualization allows the surgeon to perform complex dissection or reconstruction, ergonomic sitting reduces the stress, and wrist design removes any transfer of unwanted vibration or tremor. Surgery is minimally invasive, causing less pain, less bleeding, less recovery time, less convalescence, less blood transfusion and above all translates to less time and money. However, the instrument costs a whopping lot of money, to the tune of $1.3 million with annual maintenance of several thousand dollars, so that little more than a thousand units are sold worldwide as per available records. As it seems relevant, a time-line of the development is reproduced here for the convenience of readers, (Source: *Wikipedia*)

TIMELINE OF ROBOTIC SURGERY:

1997: A reconnection of the fallopian tubes operation was performed successfully in Cleveland using ZEUS.

1998 (May): Dr. Friedrich-Wilhelm Mohr performed the first robotically assisted heart bypass using the da Vinci Surgical System at the Leipzig Heart Centre in Germany.

1999 (September): Dr. Randall Wolf and Dr. Robert Michler performed the first robotically assisted heart bypass in USA at Ohio State University.

1999 (October): World's first surgical robotics *beating heart* coronary artery bypass graft (CABG) was performed in Canada by Dr. Douglas Boyd and Dr. Reiza Rayman using the ZEUS surgical robot.

1999 (November 22): First closed-chest beating heart cardiac hybrid revascularization procedure is performed at Health Sciences Centre (London, Ontario, Canada). At the first step of a two-step procedure, Dr. Douglas Boyd used ZEUS to perform an endoscopic single-vessel heart bypass surgery on a 55 year-old male patient's left anterior descending artery. At the next step of the procedure, William Kostuk, MD, Professor of Cardiology of the University of Western Ontario, completed an angioplasty revascularization on the patient's second occluded coronary vessel. This multi-step procedure marked one of the first integrative approaches of treating coronary diseases.

2001 (September 7): Dr. Jacques Marescaux and Dr. Michel Gagner, while in New York, used the ZEUS robotic system to remotely perform a cholecystectomy on a 68-year-old female patient who was in Strasbourg, France.

2006 (May): First unassisted robotic surgery on a 34 year old male to correct heart arrhythmia. The results were rated as better than average direct human hand skill. The machine had a database of 10,000 similar operations, and so, in the words of its designers, was "more than qualified to operate on any patient." The designers believe that robots can replace half of all surgeons within 15 years.

2008 (February): Dr. Mohan S. Gundeti of the University of Chicago Comer Children's Hospital performed the first robotic pediatric neurogenic bladder reconstruction of a 10-year-old girl.

2009 (January): Dr. Todd Tillmanns reported the results of the largest multi-institutional study on the use of the da Vinci robotic surgical system in gynecologic oncology and included learning curves for current and new users as a method to assess their acquisition of skills using the device.

2009 (January): First all-robotic kidney transplant was performed at Saint Barnabas Medical Center in Livingston, New Jersey by Dr. Stuart Geffner. The same team performed eight more fully robot-assisted kidney transplants over the next six months.

2010 (September): Eindhoven University of Technology announced the development of the SOFIE surgical system, the first surgical robot to employ force feedback.

2010 (September): First robotic operation of the femoral vasculature was performed at the University Medical Centre Ljubljana by a team led by Borut Geršak. The robot used was the first true robot, meaning it was not simply mirroring the movement of human hands, but was guided by pressing on buttons.

The question is, how technological developments at nanoscale could be associated with (or help) robotic application in surgery? Although many reputed well to do hospitals complained about recovering the cost of the instrument, it seems that sooner or later medical practice has to adapt to such modern technology. However, it is mandatory to educate the patients by intimating them the benefits in going for such a procedure with sophisticated instruments. Therefore, mass awareness and education is a necessary criterion that has to be built up over a period of time. At this stage, the developments in nanotechnology could be best integrated through the design concepts and techniques of nanomanipulators for creating more versatile robotic-hands. As outlined in Chapter II, nanomanipulators have very special designs making them capable of extremely precise handling and simultaneous 3D imaging at a level of single atoms and molecules. However, the payloads are different in both the cases even though a similarity in functions could be identified. Robotic hands are precise too but differ in maneuvering means and purposes. Technical intermingling between these two kinds of functions would initiate newer developments that might ultimately culminate into outstanding business opportunities effecting a reduction in the cost of robotic surgeries. Nevertheless, it is not easy to count chickens before they hatch, but it could be wiser to do that for good!

The biomimicry work on insect-inspired *autonomous robot fabrication* can have useful implications in this regard. It is understood that insects have much higher level of mobility than human or any land or water animals. They cover all the three dimensions, varied terrains, climb surfaces at ease and can move in any direction including doing complex motions like rotation, gyration, diving, swirling etc. Insect eyes also have much better resolution and a panoramic range of view in addition to rapid adaptability to environment. Application of robo-insects to spy on enemies is a pet idea that has made such projects a forerunner for future applications in defense.

In robo-surgery, the level of efficiency is actually limited by human efficiency, i.e. the doctor's eye to hand coordination and dexterity, if not the personal factors added to it. Whatever might be the performance capability of these super machines (e.g. *da Vinci*), the ultimate success will be dictated by the person sitting at the console and definitely there cannot be a universal performance factor. Further, there are factors of adaptation in our attitude as the takers, which need to be shifted.

Imagine however developing a robo-insect-surgeon! Such a small device can be inserted into the body through a small pierce or natural opening and then monitored remotely on a computer console to perform a surgery job through commands guided by looking at the 3D images. The device should then be deported back from the body. The robotic-hands of a robo-surgery machine will no longer be required then and those machines can as well be downsized. Special integrated laser devices on the robo-insect might be made to perform in-situ bloodless cutting or drying of the wounds. Mind that 10 millimeters is large enough to pack 1 million 10 nanometer size entities! That is the level of fabrication that nanotech can achieve and that is where nanotech fabrication might lead biomimicked robo-insect designs piloting to a new dimension in robotic surgery or medical practice in general! Machines like *da Vinci* and *Zeus* would reduce only to laptop and camera paraphernalia and therefore their cost would decrease to affordable limits to make available even at community hospitals! Remember, the first IBM computing machine, as large as to fill a room! Now you have an even better one in your cellphone! On this token, a surgeon would no more be the stethoscope hanging gentleman, but a freelance multidisciplinary laptop swinging smarty who knows not why on earth a scissor is required for surgery! As a matter of fact, as we cannot imagine now a surgical situation in the pre-anesthesia regime, the next generation would not be able to imagine bleeding a body from the outside in order to perform a procedure inside! Nanofabrication can visualize such small functional devices that have all tools of surgery integrated; still commendable in modified master-slave relation but reduced by factors at length scale! Thus, the *Future Shock* continues...

5.4.7. Nano-Robotic Therapies

The picture of robo-insect-surgeon seems somewhat like *Nanobots*. They are too distant a technology at another level of human achievement. Nanobots or nano-robots are envisaged as tiny functional molecular units much smaller than even the robo-insects and are imagined to float into the bloodstream to detect, diagnose and even treat on spot, thus preventing the body from suffering through major disorders! It performs targeted healing jobs. There is no incision required to insert them, no console to monitor them! They are programmed to perform specific tasks and then perish or flush out of the body system. Their intake is imagined as like a drug through inter-venous or injection switching mechanism. They would go inside and find the cancerous cells to destroy them only like the soldiers of Troy sent to conquer or somewhat like the *Autobots* from planet *Cybertron* to save earth from alien invasions!

This nevertheless, is imagination at the moment to build a dream that might spawn a whole range of revolutions in technologies along the way! It could be from a genetically modified bacteria or virus that already has almost all the kind of molecular level motorization mechanisms and capability of targeted delivery of genetic information. Billion years of existence through natural evolution empowered such simple organisms with all the functionality that we now understand as required for us to create. If bacteria can cause a situation that is not conducive to our health, a state that we refer as disease, then understanding this, we should also be able to modify the bacteria genetically or create an anti-bacteria or say '*acteria*' that would reinstate the situation back to normalcy!

5.5. NANOTECHNOCRATS

The world is yet to come to terms in achieving Feynman's vision of creating machines with atoms and molecules. The original thought to make nanoscale self-replicating machines by downsizing the tools of macroscopic fabrication has been realized to be unrealistic. It is understood that manufacturing with atomic precision is way harder and there are immense technical challenges to overcome whether it is a bottom-up or a top-down fabrication approach unless a revolutionary idea occurs there! There are speculations to go for a combined method where the bottom-up approach would do the fabrication of nanoscale components while the top-down would be applied to assemble them to create the functional device; such combined approaches are in fact taken up in research and several new results are being reported every now and then. Nevertheless, the state of the art in gaining a complete control to artificially steer the structure of matter at atomic and molecular level has still a long way to go!

Physics based modeling using computational methods has importance in this regard to visualize things *a priori* in order to pilot the motive towards the target. In fact, considerable progress has been done in modeling nanoscale systems and their functionality and even the self-replicating molecular machines that were coined as *nanobots* by the media. K. Eric Drexler and his associates in Nanorex are pioneers in such modeling calculations. Readers might have the details from their website: www.nanoengineer-1.net. These calculations have immense potential in figuring out how at all the imaginative future machines would function. It is indeed a daunting task to bring out such visualization considering all quantum interactions that prevail at the atomic level. The challenges of large atomistic models are enormous even in classical approximations. It is observed that the expenditure in computational resources to handle even the moderate size molecular systems from first principle using quantum mechanical descriptions is beyond what even the largest existing computational facilities are capable to provide. The results of such modeling in fact let us suppose if we could have the actual ability to manipulate matter at atomic scale, what we could have on our platter! In retrospect, these works do translate Feynman's vision into visuals in front of our eyes and help shape up the platform of our future achievements. However, judging on the immense hardships towards the practical viability of such self-replicating molecular machines, it seems that the need of the hour in computational modeling is not suggestive to have an isolated non-associative track than what the experiments are able to perform involving the available technology. Highly motivated individuals having extremely sophisticated computational techniques, or the *nanotechnocrats*, therefore should to be associated with every fundamental experimental study through interactions if not directly, might be at internet supported open platforms. It should help consolidate the newer information into knowledge for specific objective. In this regard, it might be worth here to remember what Albert Einstein remarked on the computer:

"Computers are incredibly fast, accurate and stupid. Human beings are incredibly slow, inaccurate and brilliant. Together they are powerful beyond imagination."

Therefore, modeling work is vital at every stage that at some junctures would drive the means to streamline and consolidate the knowledge and enlighten to bring focus on the target for the next phase of the journey. That indeed delineates the role of nanotechnocrats who could be a company dedicated in sophisticated computer codes to design nanoscale entities; an individual computing on certain aspects of nanofabrication or any specific nanotechnology application; or a nanoenabler computing for certain forward developments in the high profile instrument that would lead to molecular fabrication; or a researcher working on specific nanomaterial in which the calculations predict and pave the way to molecular manipulation.

The well-documented *Nanotechnology Roadmap* published in 2007 by Battelle Memorial Institute and Foresight Nanotech Institute gives a comprehensive account of all different computational methods including their applications and limitations (pages 168-171). It also highlights the possible future applications with modified and improved codes. For convenience, some important information is included; the details of which could be found in the original document, which is freely available. Distinct trends in computational approaches are identified as follows:

* *ab*-initio Molecular Mechanics (first principle)
* Molecular Dynamics (MD)
* Reactive Potentials (REBO, AIREBO, BEBOP)
* General semi-empirical methods (CNDO, INDO, MNDO, ZINDO, AMI, MP3, OM1, OM2, PM5, RM1)
* Hartee-Fock, unrestricted and open-shell (HF)
* Local Density Approximation (LDA)

* Generalized Gradient Approximations (GGA)
* Hybrid methods combining two or more, e.g.
    Car-Parrinello (CP) that combines molecular dynamics and density functional theory; Hartee-Fock with density functional theory (HF-DFT) etc.
* Moeller-Plesset Perturbation (MPn) theory
* Configuration Interaction (CI) method
* Complete Basis Set (CBS) Extrapolation method
* G2 and G3 method for thermodynamic quantities
* Multi-Configuration Self-Consistent Field (MCSCF)
* Complete Active Space Self-Consistent Field (CASSCF)
* Coupled Cluster (CC) method
* Generalized Valence Bond (GVB) calculations
* Quantum Monte Carlo (QMC) method

These software routines are instruments for modeling and designing experiments that are carried out in virtual space. Like in real experiments, these instruments also study nanoscale entities by determining the possible values of some property under certain specifications that is specified by the boundary or limiting conditions implied within the method. Like failures in real experiments due to instrumental limitations and performance restriction etc., here also the failures occur because of limitations in the program and approximations set as boundary conditions etc. With proper understanding of a certain theory and also for an evolution of certain theory, the experiments in real and virtual worlds complement each other. There may be serious disagreement on some results or issues at some point of time, even culminating into a non-compromising situation. However, this is all natural in the course of evolution of a subject and is perhaps decided by *time* in its course with regard to necessities for the advancement of civilization! Properties that computational experiments address could be anything depending on the researchers' paradigm. Nevertheless, a comprehensive listing of various properties to be followed through might be brought out from the recommendations of the *Nanotechnology Roadmap* as follows:

* Atomic/molecular ionization energies and electron affinities
* Alternative chemical reaction products and heats of formation
* Conformational energy differences
* Crystal energies and dynamics
* Dynamic friction and thermalization
* Donor/acceptor group geometries

* Electrostatic dipoles and higher multipoles and interaction energies
* Energy of reactants, products and transition state barriers
* Enzyme and catalysis binding modes
* Ground and excited state geometries and electrostatics at surfaces
* Homology models, geometric relationships
* Molecular dynamics behavior
* Molecular orbital energies, diagrams, volumes and surface areas
* Molecular self-assembly processes
* Molecular transport through proteins and pores
* Magnetic domain dynamics
* Nonlinear optical coefficients
* Optical refraction, absorption
* Protein folding and unfolding
* Solvent-based dependences of molecular properties
* Solvent-accessibility in macromolecules
* Stabilities of aggregates and ordered arrays
* Spin-spin interaction dynamics
* Transport of electron, holes and thermal energy
* Vibration and electronic transition states and energies

The *ab*-initio first principal method is accurate but it is way harder, larger and time-expensive within today's limits of computing facilities. The semi-empirical methods are comparatively faster and computationally more reasonable towards building an understanding in the subject. The *Roadmap* document also observes *"It is important to understand the limitations of each level of theory due to their approximations, as some of these limitations make the theories wholly inadequate for answering some questions."*

6. Epilogue

To conclude, it would be pertinent to quote from the *Roadmap* document once again, which says, *"Looking forward, we see both incremental payoffs and grand challenges that can be achieved through a chain of strategic objectives. Advancing from exploration, to pioneering, to full exploitation will require a great effort, but this will be a natural progression. Great rewards are already visible. They merit a commensurate investment."*

The chapter began by noting our practice of planning: it has to be a daily habit in order that we can wrap up with the 'Nano-Net.'

I tried to reiterate through all discussions that,

1. It is mandatory to change the coin by which investment is measured conventionally.

2. It is required to keep faith on the promises that is given by science for the future and *speculate daringly* even though the path is going to be tumultuous.

3. A stake is not what is estimated by the data of *stock*, but also the input to make our planet more advanced and a better place to live.

4. No dream is here for quick money but there are dreams to keep us awake forever.

5. Ethics would always oppose at any time, being built on tradition and belief. The need for progress is to debate under the new light of science and redefine ethics.

6. It says, money decides in modern civilization. That is indeed short-sightedness for the new future on horizon!

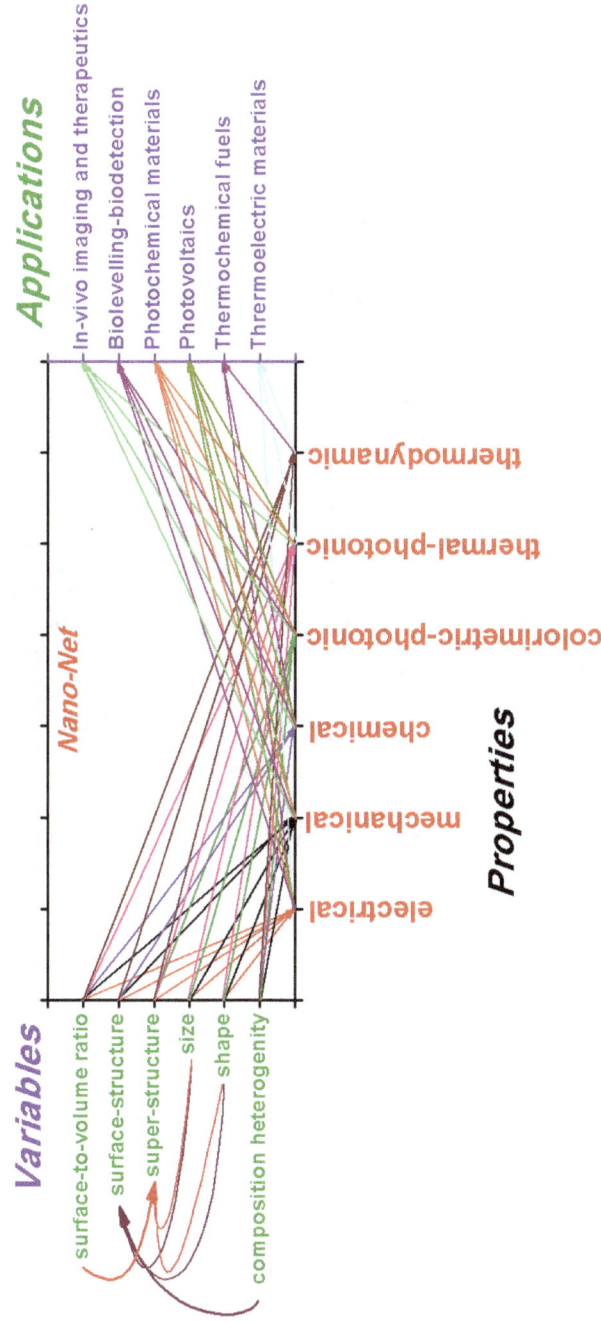

Fig. 6. The 'Nano-Net': shows the interrelation of the fundamental properties with possible applications and the variables that enable to tune them – an exclusive speciality in the domain of nanotechnology. (Source: James Heath, *Nanoscience and Technology, A Collection of Reviews from Nature Journals, Ed. Peter Rodgers.* World Scientific, 2010.)

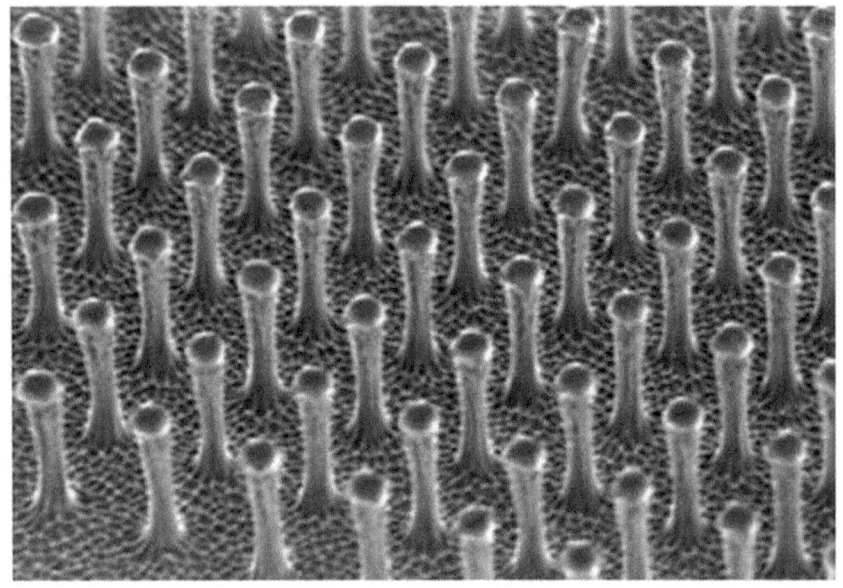

Fig. 7. The Gecko Tape under microscope (Courtesy: http://www.newscientist.com)

Steps to form nanostructure using frustule

1. Place a chosen frustule on suitable wafer

2. Deposit metal atoms by evaporation

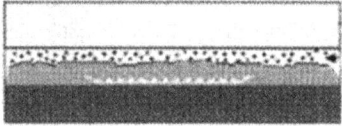

3. On completion, Glue a glass plate on top

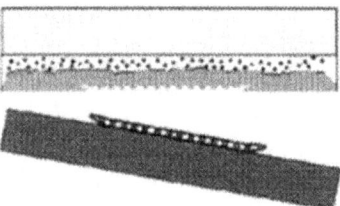

4. Remove the template now to obtain a replica

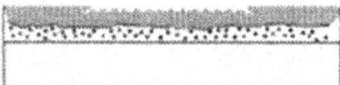

Fig. 8. Assorted image of various Diatom Frustules and its use to create metal nanostructures is shown in schematics. (Image reproduced from *New Journal of Chemistry*, Vol. 30, (2008), p. 908 with permission of the Royal Society of Chemistry. Frustule image courtesy: http://uoregongk12.wikispaces.com)

CHAPTER IV

The Roadmap

In order to be an immaculate member of a flock
of sheep one must, above all, be a sheep.
—Albert Einstein

It would be worth highlighting the *Nanotechnology Roadmap* 2007 authored by BATTELLE, not only for the sake of redirecting interested readers to look into more details but also for easing that document for non-technical readers. Actually, the Roadmap document presents huge amount of matters more specifically like in a legal report than like what is commonly written for papers in a scientific journal. A journal article normally is put up in a logic-driven inductive style. However, a Roadmap document needs to encompass very huge amount of diverse information and it is indeed a daunting task to reclassify all different activities in terms of their significance for nanotechnology applications. A Roadmap document obviously has a special character in its presentation. To streamline the collective knowledge, such a pioneering pamphlet is extremely important and appreciatory in figuring out the future course of our journey in nanotechnology. By itself, this document redefines the entire gamut of activities. However, the apparent nonconventional classification initially appears unfamiliar and gives a jitter even to a science practitioner! The cloud gets clear subsequently as one sails through the examples and eventually, the entire matter seems comprehensive. Nevertheless, the use of the adjectives *atomically precise* is noteworthy in this report compared to that published by IWGN in 2004. It appears as an overemphasis to something that is already implied and does not practically help to erect any fence separating the domain of nanotechnology, which anyway, is nonessential. The matter becomes more user (memory) friendly by keeping these two words sensibly inherent.

The following different research activities have been suggested as enabling nanotechnology currently:

* Structural DNA nanotechnology
* Scanning probe manipulation
* Protein design
* Macromolecular self-assembly
* Nanoparticle synthesis
* Nanolithography
* Organic synthesis
* Biotechnology and molecular biology
* Surface science
* Molecular imaging

The entire range of activities is reclassified under the following nine broader headings:

* Atomic Precision in Manufacturing
* Components and Devices
* Motors and Actuators
* Motion Control
* Molecular Processing
* Computational Modeling
* Signal Transduction
* Energy Manipulation
* Photonics

Any activity that contributes in the fundamental understanding or application in nanotechnology may be identified to belong to one or more of the categories in this grouping. However, these are not mandatory in practice but important to exercise a discipline and streamlining the vast amount of information being accumulated on a daily basis. Notwithstanding, the different new technical terminologies defined in the document, e.g. APT, APM, APPN etc., essentially reformulates the knowledge in an ordered manner. The abbreviations might also be read as follows: Atomic [A] Precision [P] of manipulation in Technology [T] to Manufacture [M] systems and devices that are Productive (or perform machine level production) at Nanometer scale —Productive Nanosystems (PN).

The timeframe of the roadmap is given till 2030. Six years have passed from its first publication. To quintessentially reassess and reset the yardstick, it might be useful to tabulate the essential points laid in the Roadmap document for a quick ready reference:

Nanotechnology: Atomic Precision in Manufacturing

Agenda for research
"To develop complex systems, efforts must be coordinated so as to develop all the parts they require. This entails selecting and refining objectives, determining requirements, considering options for meeting them, and thereby identifying research directions that are more likely to produce results of great value."

1. Scanning probe based manufacturing by tip manipulation, array formation from atomic and nanocluster deposition → solid building, scalable and scaled array systems
2. Nanocluster synthesis, deposition and matrix isolated assembly → modular fabrication
3. Organic synthesis, organometallics → self-assembly of composite nanosystems
4. Protein engineering and Ribosome as productive nanosystem → Bio-based synthesis
5. Structural DNA design → modular molecular composite nanosystem
6. Polymerase → productive nanosystems
7. Nanotube, nanocrystals, nanocomposites → device fabrication

Components and Devices *i.e. passive and active components of kinds that may prove useful in implementing atomically precise functional nanosystems*	Structural Components →	** Modular oligomers* ** Surfaces* ** Sheets and fibers* ** Dendrimers* ** Biological nanoparticles* ** Ceramic nanocrystals* ** Metallic nanocrystals* ** Graphene components* ** Inorganic nanotubes* ** Semiconductor and metallic nanowires*
	Motors and Actuators →	** Biological* ** Synthetic* ** Macroscopic*
	Motion Control →	** Molecular bearings: sigma-bond* ** gears, clutches hinges: H-bond; dative-bond*

Molecule Processing →	* *Catalysis* * *Enzymes* * *Atomic ledges, kinks and ad-atoms* * *Active tips* * *Filtration membrane pores* * *Soluble or volatile precursors*
Computation Modeling →	* *Logic operations* * *Memory* * *Mechanical computation components* * *Quantum dot cellular automata* * *Coherent quantum computation* * *Signal transmission*
Signal Transduction →	* *Optical to mechanical, electrical and the reverse* * *Chemical to mechanical, optical, electrical, magnetic, and the reverse* * *Molecular sensing*
Energy Manipulation →	* *Storage* * *Conversion*
Photonics →	* *Linear optical components* * *Band gap materials* * *Meta-materials* * *Nonlinear transmission* * *Nonlinear harmonic generation* * *Controllable absorption, phase modulation*

Systems and Frameworks *i.e. systems and subsystems that can serve roles in nanosystems engineering and its applications*	Structural Frameworks →	* *Using scanning probes (TIP)* * *Based on self-assembly crystal, DNA, protein components* * *By modular molecular composites* * *DNA engineering, Protein engineering, Special structures*
	Machinery, Active Positioning Systems → Mechanisms that drive the motion of nanoscale compo-nents	* *Protect active surfaces* * *Position catalytic sites* * *Modulate transfer of energy* * *Direct molecular reactivity* * *Construct systems not possible by self-assembly* * *Move nanosystems to specific location*
	Productive Nanosystems → that can be used to make any of a wide variety of structures under programmable control	* *Natural → Ribosomes, RNA polymerase, Reverse transcriptase, DNA polymerase, RNA to RNA polymerase* * *Synthetic → Ribosome class artificial systems, 2D and 3D polymeric component builders, component construction via self-assembly and direct manipulation*
	Systems for Application Areas → Potential areas of application for nanosystems, at product level as well as subsystem level	* *Information processing add-ons to advanced semiconductor systems* * *Full computational systems* * *Information-oriented optical systems* * *Medical systems: discussed previously in detail*

		* Energy conservation systems *→ Photovoltaic,* *Photochemical, Electro-* *chemical, Fuel Cells, Batteries*
Fabrication and Synthesis Methods *i.e. Techniques for fabricating atomically precise components* *AND* *Technologies facilitating the development or application of atomically precise systems.*	Organic Synthesis →	*To create finite set of functional components and building blocks on the order of 10 to 100 atoms in size*
	Atomically Precise Self-assembly →	** DNA self-assembly* ** Protein self-assembly* ** Shape programmable oligomers e.g. bis-peptides* ** Chemical self-assembly*
	→ Scanning Probe based Fabrication: i.e. Fabrication using STM, AFM, SPM, FFM etc.	***Mechanosynthesis*** ** Structures by surface atom manipulation* ** Array or structures from weekly bound deposit atoms or nanoclusters* ** Patterned atomic layer epitaxy (PALE)* ** Placement based synthesis*
	Hybrid Fabrication → Different methods in combination making nanoscale components	** Graphene-based structures e.g. C_{60}, C_{70}, Carbon Nanotube and Graphene* ** Semiconductor nanocrystals, e.g. Q-dots* ** Metal nanocluster, and other passivated clusters*

	Imprecise Fabrication → Top-down approaches: Lithography <45 nm node	*Optical* → *30 nm* *Electron-beam* → *5 nm* *Focused ion-beam* → *few nm* *Helium-beam* → *0.25 nm* *Nanoimprint* → *<50nm* *Dip Pen Nanolithography* *Self-assembled monolayer*
	Methods	*Properties*
Designing **and** **Modeling** *i.e. Guides for fabrication by providing a theoretical framework for generating and testing structures, devices and entire systems by means of computational experiments*	* ab-initio *Molecular Dynamics * Reactive Potentials * Semi-Empirical * Hartee-Fock * Local Density Approx. * Generalized Gradient Approx. * Hybrid methods * Moeller-Plesset Perturbation * Configuration Interaction * Complete Basis Set * G2 and G3 * Multi-Configuration self-Consistent Field * Complete Active Space self-Consistent Field * Coupled Cluster * Generalized Valence Bond * Quantum Monte Carlo	*Ionization energies* *Electron affinities* *Chemical reaction products* *Heats of formation* *Conformational energy difference* *Crystal energies and dynamics* *Dynamic friction* *Thermalization* *Donor/acceptor group geometries* *Electrostatic dipoles and higher multipoles* *Interaction energies* *Energy of reactants, products and transition state barriers* *Enzyme and catalysis binding modes* *Ground and excited state geometries* *Electrostatics at surfaces* *Homology, geometric relationships* *Molecular behavior* *Orbital energies, diagrams,* *Volume, surface area*

		* *Molecular self-assembly processes* * *Molecular transport through proteins and pores* * *Magnetic domain dynamics* * *Nonlinear optical coefficients* * *Optical refraction, absorption* * *Protein folding and unfolding* * *Solvent dependence of properties* * *Solvent-accessibility in macromolecules* * *Stabilities of aggregates* * *Ordered arrays* * *Spin-spin interaction* * *Transport of electron, holes and energy* * *Vibration and electronic transition states and energies*
	Instrument/Technique	***Information Gain***
Characterization *Characterization of structure and functional properties is critical to developing components and systems on any scale, including the nanoscale. Improvement in breadth, robustness and precision of characterization tools are important because they can speed acquisition and improve the quality of information about products.*	Microscopes: scanning tunneling (STM), atomic force (AFM), scanning probe (SPM), friction force (FFM)	*Atomic scale measurements, Manipulation and Fabrication*
	Scanning electronic microscope (SEM)	*Particle size, Morphology, Segregation*
	Fast Ion Bombardment (FIB)	*Topography, 3D Composition, Crystallography*
	Transmission Electron Microscope (TEM)	*Phase, Structure, Composition, Size and Chemical state*
Some examples →	X-Ray Diffraction (XRD)	*Crystal structure, Lattice constant, Molecular geometry*

Cryo-Electron Tomography	*3-D Structure*
Scanning Helium ion microscopy	*Topography, Chemical contrast*
Mass spectrometry (MS) Tandem MS-MS and Secondary Ion Mass Spectrometry (SIMS)	*Nanocluster and particle size distribution, Electronic and geometric structural information, Surface composition, Impurity, Atomic and Isotopic contents*
Scanning Auger Microscopy	*Size, shape, surface and 3-D composition*
X-ray Photoelectron spectroscopy	*Average surface composition and Chemical state*
Small angle X-ray and Neutron scattering	*Local chemical environment, Geometry, Size and Surface clustering*
X-ray absorption	*Oxidation state, Salvation structure*
Proton induced X-ray emission	*Elemental composition*
Tetrahertz spectroscopy	*Low frequency vibration modes, Inter and Intra molecular interactions*
Raman spectroscopy	*Energy of vibration modes, Molecular conformation, Molecular interactions*
Fourier Transform Infrared Spectroscopy FTIR and Near Infrared Spectroscopy (NIR)	*Vibration energies, Molecular interactions*

UV and Visible spectroscopy	*Electronic states, Optical properties, Photochemistry*
Neutron scattering spectroscopy	*Phonon modes, Molecular geometry via selective deuteration, Magnetic structure*
Nuclear Magnetic Resonance (NMR)	*Magnetic environment, Molecular structure, Secondary structure, Chemical environment*
Dynamic Laser scattering	*Particle size distribution till 5 nm*
Phase analysis light scattering	*Charge state of particle*
Disc centrifuge photo sedimentation	*Particle size distribution till 3 nm*

This table attempts to give an essence of different sections and subsections of the 176 page *Roadmap* document. The purpose of making such a tabulation is to assist the reader to quickly follow through the original document in order to find the details of the topic of interest. Effort has been put to make the tabulation precise without repetition. However, it only refers back the original document, which is highly important and a class by itself. The motto here is to point out the necessity of a patient reading of something that encounters the valuable observations of collective wisdom, for example by saying,

> "*The R&D effort as a whole must closely interweave developments in fundamental understanding of nanoscale properties, new materials synthesis methodologies, new manufacturing techniques, new characterization and control techniques, and new modeling tools. Progress in nanosystems development requires iterative cycles of design, modeling, fabrication, and characterization. All these steps are necessary, and each step and field of application presents a rich and diverse set of multi-disciplinary challenges.*"

The document also enumerates,

"Some research paths lead toward ordinary applications, but other paths lead toward strategic objectives that are broadly enabling, objectives that can open many paths and create new fields. These paths are the focus of this roadmap. They demand further exploration."

In order to conclude, it might be pertinent to put up certain observations commensurate with the general nature of growth of scientific knowledge. It needs the conditions of growth of a tree, like freedom of thought as sunlight; freedom of information exchange as air; freedom of use of existing knowledge as soil; freedom to express and share as water. Growth should be unconditional and not under the burden of any parameter so that all the branches can develop in any direction which *nature* permits; new leaf can come out every day that enriches through the nutrition of research and slowly become mature as its color turns deeper and it becomes heavier; it is then capable to give birth of further new leaves! True, that scientific research demands money to support but that should be necessity driven and not profit driven. Science exchange is to enrich the soil of existing knowledge base and not for wrapping up in fancy reports with high price tags that have presently became fashionable in some quarters creating the hype for nanotechnology as a lucrative business and organizing meetings in 7-star hotels with costly registration fees! It should now be told in straight and simple way that such practice are impediments to normal and healthy growth of a subject that should not be encouraged at least by people who are engaged in the practice of science. There is no marriage of business with science, business is essentially a cofactor in the growth of science but it cannot be the enzyme to drive the reaction. Business is always associated with science but science cannot be a commodity for business. Free development only can create new avenues to do new business!

In this regard, something might be learned from the study of the growth of the internet (world wide web) around the globe. There is actually nothing in science that developed so rapidly like the internet and gained such an extensive use within a very short period of time! Once again, thanks go to Kevin Kelly who perhaps described it best (in TED). It is amazing to believe now that there are 100 billion clicks per day and 55 trillion links between the web pages of the world! The important thing to learn from this phenomenal happening is the *power of freedom in growth*. Internet has grown freely without the bindings of any particular subject or the rulings of any country or the reign of any philosophical dictum. Of course, free growth has generated some vices as the brainchild of some unscrupulous minds. But, at any point of time the positive applications always supersede in number and have been superior in quality than any negative utility. In predicting the scenario of the internet for next 5,000 days, Kelly identifies three attributes. One, the internet will be given a body, i.e. internationally agreed recognition of standard and reliability; two, restructuring of its architecture into a new paradigm of performance in information exchange; and three, people will become completely co-dependent upon it, i.e. it is an indispensable part of us and we are an inevitable part of it! All these characteristics could be inherited for the growth of nanotechnology in next 5,000 days! Therefore,

* What we need to do: share data as, Kevin Kelly says, *"To share is to gain."*
* What we should be: co-dependent, by default that would mean to be absolutely transparent in practice.
* What we would become: ubiquitous, that would inculcate a new culture within the existing setup!

CHAPTER V

Tube, Titania, Textile and Nano

> Innovation is not the product of logical thought, even though
> the final product is tied to a logical structure.
> —Albert Einstein

Innovation of certain novel nanostructures actually should be considered as the blessings in disguise in the journey of attaining the final goal of making atomically engineered precise nanoscale devices which Feynman envisioned. However, these nanostructures are not made by using engineering tools at atomic scale, rather obtained as formations under specific physical conditions. C_{60}, the Buckminsterfullerene and Carbon Nanotube (CNT) are to be considered in this category. Interestingly, carbon is the element for such special creations and is also the core element in all biological molecules!

C_{60} is very well known now though it did not find a wide range of applications as CNT. There are other interesting cage structures of carbon atoms as well which are discovered after C_{60} in fact, many a times during the course of post-treatment experiments, e.g. C_{70}, Carbon-Onion etc. These are all very interesting examples which depict the reality on how the graphene sheet could loop on itself giving rise to enclosed formations. Descriptions of their structures and some properties are published in many literatures of which some are quoted in the bibliography. Among these, CNT, unanimous as *nanotube*, has evolved as the most successful one in terms of applications that it might as well be designated now as the *material of the future* or the *Diamond of the Nanoworld*.

1. Carbon Nanotube

Actually, nanotube is not new! It was used even in our ancient civilization, however, was not familiar in this name. In 2006, researchers found that the steel used in 17th century to make the famous *Damascus Sabre* in fact contained nanotubes. That was essentially the reason for the extraordinary strength of those blades. European civilization was not familiar with this steel. In fact, the material of these fine Syrian craftsmanship used to be *Wootz* steel manufactured in ancient India, from which the swords were to be forged directly. This material had indeed been a very precious commodity for the kings in ancient times. Indian *Wootz* production has been reported to be a very special metallurgical procedure that happened to use the ore only from particular mines and special quality *Avaram* wood with selective leaves of the medicinal *giant rubber bush* plant. Thermal cycling and periodic forging is reported to cause the formation of cementite nanowires and coarse cementite particles, giving rise to the extraordinary toughness. Perhaps, the blade-smiths were not aware about the alloying that happens with these ingredients but certainly knew about the mechanical properties of the end product. Thus, though unseen, nanotube was serving some purpose for human society for centuries together as *nature's* hidden boon before being discovered under the scholarly pursuit.

A timeline for this rediscovery or an account of the unfolding in modern era might be listed as follows:

1952: *The Soviet Journal of Physical Chemistry* reported the image of the formation small cylindrical structures of carbon having approximately a diameter of about 50 nm. These tubules and bundles were called '*barrelenes*.' The aspect ratio of these nanostructures was much smaller and somewhat like the elongated fullerenes, which was discovered much later.

1960: Carbon whiskers were reported by Roger Bacon from Union Carbide Corporation, Cleveland, Ohio USA. These ribbon-like structures were synthesized by DC arc-discharge at very high buffer-gas (argon) pressure and at an elevated temperature. They are from fraction of a micron to over 5 micron in diameter and up to 3 cm in length. The XRD pattern revealed that they are more like a scroll of graphene sheets creating what is known today as *parchment multiwall nanotubes*. The high tensile strength and Young's modulus indeed proves them to be the multiwall nanotubes.

1975-1978: M. Endo did a PhD thesis at University of Orleans, France in 1975 on the studies of small diameter carbon filaments. A variety of conditions of their synthesis have been reported. Publication in 1976 by Oberlin, Endo and Koyama clearly showed hollow carbon fiber using the vapor-growth technique. In these finer nanometer diameter filaments, evidence of tubes with single graphene wall is also found, but they did not call these entities nanotubes then! M. Endo has another thesis at Nagoya University, Japan in 1978.

1979: Special carbon structures in the name of carbon fibers were reported at the 14th Biennial Conference of Carbon. These fibers were as well synthesized by arc-discharge but at atmospheric pressure with nitrogen as the buffer gas instead of argon. The fibers turned out to be multi-wall nanotubes having 5-16 nm diameters. The largest average spacing between two carbon atoms has been determined to be 0.35 nm from electron diffraction studies.

1980s: Soviet scientists published the formation of tubular crystals of multi-layer carbon by rolling graphene layers into cylinders. TEM images were published along with XRD data suggesting the possibility of many different structural formations like circular, spiral and helical etc. Under modern descriptions, these arrangements conform to *armchair* and *chiral* nanotubes respectively. A US patent was issued in 1987 to Hyperion Catalysis for production of *cylindrical carbon fibrils*. These were micron length multiwall nanotubes having 3-70 nm diameter.

1991: Ijima reported for the first time on the synthesis of multiwall nanotubes by arc-discharge of graphite rods. Interestingly, this did not create much of an attention at that time. In the same year however, an interesting calculation was also published quite independently by Mintmire et al., depicting extraordinary transport properties of a cylindrical construction made up of one layer of carbon atoms bonded together in honeycomb lattice. Perhaps, this was the beginning of single wall nanotube (SWNT) though they called the calculated structures as *fullerene tubules*. This theoretical study was taken seriously to make tubular structures having a wall made of single layer of bonded carbon atoms. The excitement was followed quickly by Ijima's second discovery of such a structure in 1993 and an independent work by Bethune et al. at the same time. Two publications in the prestigious British journal *Nature* in the same year on the same subject created quite a global sensation immediately to be followed up through several measurements on their conducting properties. The important thing to note is that both reported the use of transition metal as catalyst in arc-discharge synthesis, which was never done before in the case of fullerene or fullerene-type formations. Thus, a precious entity was born in the nanoworld through an intellectual pursuit that has actually been going over a long period on collective conscious. The uniqueness of this entity made it a class by itself!

Certainly, Ijima's 1991 report in the journal *Nature* was basically on a scroll-in type multiwall nanotube though, importantly it created a broader awareness. The theory paper in *Physical Review Letters* in 1992 has indeed helped the growth of an interest within the scientific community. In fact, the recent astonishing results almost on a daily basis on different studies on nanotubes suggest that it is no longer important to keep the discovery issue contentious. Obviously, CNT is not discovered suddenly at certain place and time! On the contrary, it is actually the culmination of collective wisdom accumulated over decades that revealed the secret existence of this wonderful entity in *nature*!

1.1. STRUCTURE

Hence, we learned how to isolate one or a few layers of graphene from graphite and construct a spherical or a cylindrical structure. Overall, these are chemico-physical synthesis. The structures are essentially generated in order to relax the stress out of the system which is exerted through the external conditions of synthesis. In this action, an end-on rolling produces SWNT or concentric MWNT and an end-off roll-in-around result into scroll or parchment type formation. However, the methods of fabrication are yet to be completely standardized, e.g. like in metallurgical process. It is required to discover procedures for batch production of particular type of products, e.g. all metallic 5 nm diameter nanotubes or there should be standardized after-treatment separation techniques of various nanotubes in terms of size and character. Recently, some success is achieved in this regard and will be described subsequently. Nevertheless, the synthesis can very well be grouped in the category of *Atomically Precise Fabrication*. It is therefore extremely important to know these tiny tubular structures in atomic scale so that the arrangement of carbon atoms creating such beautiful entity could be appreciated.

A graphene sheet is the unit that makes graphite where they are stacked together by van der Waals force. It is a two dimensional layer of carbon atoms so arranged that if you draw lines to join each one with the other, you would end up having a geometrical pattern that resembles to what the nest of honey bees looks like. All identical hexagons are juxtaposed wherein every corner has a carbon atom. A representative drawing is shown in Fig. 10 for readers' convenience. Actually, physicists always love to have a quantitative description of everything in *nature!* There is specific description of such a structure that uses certain mathematical definitions.

To understand the matter but avoid scientific jargon, it is best to inscribe a number coordinate to each corner where the carbon atoms are supposed to be. The origin (0, 0) is an arbitrary choice to create a reference of the construct. It is interesting to note that each such point is connected by three lines and in case of regular hexagons these lines are 120° apart from each other. Therefore, each carbon atom is connected with three others unlike, e.g. say in methane (CH_4) where one carbon atom is connected with four hydrogen atoms. Such a connectivity, which creates three immediate neighbors to each, is referred as sp^2 because of the orbital hybridization scheme. Nothing to worry if you don't know what that specifically means in order to follow what happens when a graphene sheet is rolled over to construct a tube! If every corner is designated by a pair of numbers as shown in Fig. 10, it would immediately follow that the points with same digits as coordinates like (1, 1), (2, 2), (3, 3) etc. will fall along a straight line 30° apart from any of the zero-lines, i.e., from both the (x, 0) or (0, y) lines. By dint of the fact, there can only be three typical situations when the sheet is rolled over which can be directly verified simply by drawing Fig. 10 on a transparent paper and then rolling it over to construct a tube;

* Zero degree rollover, i.e. around any of the zero-lines. In this tube, one carbon atom apparently shows at the center of any hexagon when looked across. This is the type called Zigzag tubes, of which a drawing is shown at the top of Fig. 10.
* If the rolling takes place along the 30° diagonal line, i.e. the line having same digits for both coordinates, the tube looks as if a see-through one. Here, hexagons exactly overlap. This type of tube is called the Armchair tube, the one shown at the bottom of Fig. 10.
* When rollover takes place in any other arbitrary angle between 0 and 30 degrees, the tubes are called Chiral. This type is shown at the left side in Fig. 10.

The nominal diameter of a certain tube specified by certain coordinate can be calculated from the formula,

$$d = \frac{0.246}{\pi} \sqrt{(x^2 + xy + y^2)}$$

The most interesting fact is that the electrical transport property of these three types of tubes varies in surprising proportion. Armchair tubes turns out to be metallic whereas both the zigzag and chiral ones are mostly semiconducting. The band gap can vary from 0 to 2 eV depending on how the carbon atoms are mutually arranged with respect to each other as a result of the typical roll over. CNT, in which there is only single layer of carbon atoms or a single wall, are referred as Single Wall Nano Tube or SWNT. Most of their properties change, at least theoretically, with the (x, y) values and the dependence is non-monotonic. For example, a (9, 0) zigzag tube has a diameter of 0.715 nm as against the (5, 5) armchair tube that would have 0.688 nm. It reveals that the physical character of materials does depend on the shape and size, which, importantly, can act as the handler to tune up the properties for specific purpose; a phenomenon not experienced in materials science before! This also certifies the uniqueness of materials in the domain of nano as explained earlier in Chapter-III. In MWNT, it is quite possible to have the different types of tubes forming one concentric tube. In this, the double wall nanotubes are found to be very typical. Their properties almost resemble that of SWNT but they are more resistive to chemicals. The scroll MWNTs are geometrically more difficult but are mostly expected as chiral tubes. As such, MWNTs are predicted to be interesting for their use in creating nanoscale robotic functions and as a matter of fact, could be a stepping stone to create the ultimate molecular machines of Feynman!

In this connection, it is pertinent to have a comparative understanding of the structures of nanotube with respect to fullerene that is a closed cage of carbon atoms, which can be seen as the 0D (or 0-dimensional) entity, if graphene is 2D and graphite is 3D material. Actually, a complete hexagonal net cannot construct a closed structure. Pentagonal faces are required to close it up into a cage configuration. The soccer ball like fullerene has 12 pentagonal faces with 20 hexagonal ones. On the other hand, an all-pentagonal infinite planer net is not possible to construct. In fact, the all-pentagonal closed cage has 20 vertices that is one of the isomeric structures of C_{20}. There is in fact a simple set of rules given by famous mathematician Euler in this regard where,

$$no\ of\ vertices, n_v = \frac{1}{3}(5P + 6H)$$

$$no\ of\ edges, n_e = \frac{1}{2}(5P + 6H)$$

Here P and H stands for the number of pentagons and hexagons and the number of faces is given by the sum of the number of pentagons and hexagons. In general, these structures are known for a long time as Polyhedral Cages. Lord Kelvin, in 1887, suggested the 24 vertex cage ($12P+2H$) known as *tetrakaidecahedra* in order to explain the three dimensional shape of cells in the tissue which might give the maximum dense packing (overlapped onto the square faces). These structures, referred to as Buckyball Water Clusters (BWC), are also used to explain the structures of large water clusters, believed to be the constituent of clouds in the upper atmosphere. It is interesting to observe here that the smallest closed cage contains 12 pentagons and have 20 vertices with 20 carbon atoms placed in 3-coordination. As the size of the cage increases, the growth happens at the cost of addition of hexagons in which each hexagon adds up two more vertices so that C_{60} turns out to be a structure having 20 hexagons and 12 pentagons. Similarly, C_{70} contains 25 hexagons with 12 pentagons. It is indeed interesting to decipher how the stress introduced into the system could initiate a competition between caging and rolling in the same graphitic sheet!

In this regard, the recent interesting development on freestanding graphitic layer is worth mentioning. An independent stand-alone monolayer of carbon atoms in hexagonal network forming an ideal honeycomb lattice is called graphene. Actually, famous physicist Landau and Peierls, 70 years ago showed that an ideal 2D crystal of atoms should be thermodynamically unstable and hence, is not supposed to exist. However, very recently, micron size ideal 2D lattice of carbon atoms has been synthesized on top of silicon-carbide (SiC) surface by ultra-high vacuum evaporation. The near-ideal ones are achieved through several techniques for example, physical drawing from well defined graphite surface and also through the chemical synthesis of reduced graphite oxide (RGO) flakes in which, portions of 2D networks are linked through 3D oxygenated zones. There is a huge interest grown on such 2D layers of carbon atoms because of very serious theoretical insight on graphene which is expected to exhibit important quantum mechanical phenomenon, and even the high-energy particle physics on a table-top system! In fact, the theoretical studies have been around for quite sometime now which made such a system to be known as an *academic material*. Theoretically, such an open system of carbon atoms should have charge carriers as massless Dirac fermions! The interaction of moving electrons with the periodic potential of honeycomb lattice, gives rise to new quasi-particles which obey the Dirac's equation. These quasi-particles, almost moving at the speed of light, are known as massless Dirac fermions which could also be viewed as the neutrinos that have acquired an electronic charge. It is quite a remarkable phenomenon because in condensed matter physics, Schrödinger equation is the most important and omnipresent! From the point of view of electronic properties, graphene is a zero-gap semiconductor in which, the charge carriers can travel over several interatomic distances without suffering any scattering; thus makes the possibility of ballistic transport of charge carriers, into a reality. Such carrier transport can give rise to new electronics. Recently, graphene has been found to exhibit a behaviour like quantum dots or as an array of quantum dots in the form of RGO. Such novel predictions and findings has termed graphene as a wonder-material and initiated a new *gold-rush* with anticipation of revolutionizing the face of modern electronics, vis-à-vis our life on earth!

1.2. SYNTHESIS

The most common method of synthesis of CNT is arc discharge wherein a very large current (about 80-100 amps) is passed through two graphite electrodes and a discharge is created at critical buffer gas pressure. The shoots produced in the process is collected, typically in a cleverly designed collector placed at a convenient proximity that catches almost all the materials which are then separated, conventionally using a centrifuge. This kind of source is well known as Smoke Source. A typical arrangement is shown in Fig. 11. This kind of source is easy to build except that it needs a high stability large current power supply. Over the years, the synthesis technique has improved and it is found that use of certain transition metals as catalyst results into more specific formations.

Laser Ablation (LA) and Chemical Vapour Deposition (CVD) have turned out to be more useful for a more regular growth of carbon nanotubes. In fact, different other sources of carbon than graphite, even camphor, have been found to be successful in making good quality nanotubes. The main difference between these two methods is in the technique of graphite evaporation. In the laser ablation process, a small quantity from a solid graphite target is evaporated at each laser shot and the vapor is then let to condense continuously to form the shoot. In vapor transport method, the vapor is transported to a zone of colder temperature maintained over a suitable distance wherein it condenses, or otherwise, the vapor is allowed to condense on a colder substrate to form the shoot. However, quality wise, laser ablation is a better choice for synthesizing SWNT because almost 70% of the shoot produced in this method contains SWNT compared to arc-discharge method in which on average 30% of the yield contains both single and multiwall nanotubes in a mix. The diameter of the tubes can also be controlled to some extent by controlling the reaction temperature. Cobalt and Nickel or a mixture of them is normally used as good catalysts in the growth process.

Chemical vapor transport has been a well known technique to synthesize crystals and have been used extensively both at laboratories and in industries as well. First CVD synthesis was reported in 1993 following Ijima's publication on arc discharge method. CVD has been found more suitable for large quantity commercial production and the price per unit ratio in this process turns out to be much lower compared to any other methods. One more advantage in this process is the possibility of growing tubes on a desired substrate. The use of metal nanoparticles is found to produce a much larger yield. However, the detailed descriptions of different synthesis procedures with comparative accounts of respective merits and demerits, is out of the context of this book and are available in plenty of literature for anybody to choose his suitable method to make CNT.

Catalytic decomposition of any suitable hydrocarbon as the starting material can also lead to the production of CNT, be it alcohol or camphor, if the control parameters can be well optimized. All these parent materials mostly produce MWNT and the important concern is which one would become the most cost effective for mass production without a compromise in environmental and safety aspects. For technological applications, the need of the hour is to find well designed controlled growth of single type of nanotubes in large scale or a very user friendly procedure of separating the different types of nanotubes out of the product mixture. It is not surprising to understand that carbon atoms in an sp^2 bonding situation are energetically and thermodynamically most stable in a hexagonal network, but under a controlled stress it might minimize the internal energy by coiling up a portion into cylindrical formations. The chemistry of this conversion from an sp^3 bonded organic material into an sp^2 bonded structure should be very intricate with intermediate steps assisted through catalysts, which will slowly unfold in the course of many different future studies.

One recent study at Rice University is remarkable in this regard. They have succeeded in performing the seeded growth of a single type of nanotube. This is actually a very specific technique wherein small cut pieces of a particular nanotube are used as the seeds to sprout new SWNT. The idea to use nanotubes themselves as the precursor to further growth is very novel but in practice, it is a nontrivial job because cutting a nanotube for making seeds is technically difficult. Lately, the group has reported the growth of long armchair nanotubes using such techniques. There is large hope that this technique would help in synthesizing the desired type of nanotubes in large quantities in near future. However, the complexity involved in the method might be a deterrent to cost effectiveness and replacing today's electrical grids with nanowire grids is still a distant reality! Nevertheless, better solutions are not necessarily the cheaper ones!

1.3 PROPERTIES

Overwhelming amount of studies is being done since the discovery to know various properties of carbon nanotube for application in technology. In accordance with the intent of this book, it would be pertinent to enlist different achievements instead of detailing a description of all the research findings that invented them. The properties actually manifest the intrinsic nature of this nanostructure and qualify it. As per the graph in Chapter III (Fig. 6), they form the base for diverse technological applications. The following table is a subjective listing summarizing the most important properties observed so far:

Important Properties of Carbon Nanotube

Mechanical	1. The strongest and stiffest material discovered so far. Tensile strength of MWNT is measured to be 63-150 Giga Pascal (GPa), i.e. a 1 mm diameter rope made out of CNT would support a load of more than 6 tons! That translates into the strength 100 times more than steel and nearly 40 times more than Kevlar. The maximum tensile strength measured in an armchair metallic single wall nanotube is more than 126 GPa. This data is actually the genesis of the fable of making a lift to reach above the cloud. 2. Excessive load may cause permanent plastic deformation of the tubes. 3. However, the mechanical strength is along the tube axis only. Whereas, it is much softer along the radial direction. This way it is still not superior to steel and Kevlar for general applications.
Electrical	CNT has unique electronic structure that is pivotal to the specific atomic arrangement due to the particular geometry. However, this is yet to be made tunable.

In general, CNT are semiconductors, i.e. all chiral tubes are semiconducting however,

* For any (n, m) tube if n = m, then they are metallic i.e. all armchair tubes are metallic and,

* If (n—m) is a multiple of 3, they are narrow-band semiconductors i.e. zigzag tubes can be either metallic or semiconducting.

* There is strong influence of the geometry, especially the curvature that creates many exceptions to the rules in transport property.

* The current density is found to be even 1000 times more than copper, typically in the range of $(3-5)*10^9$ A/cm^2.

* They exhibit superconductivity in the range of liquid helium temperature, however, MWNT that has interconnected inner shells are found to be superconducting at and below 12 K temperature. This enhancement in super-conducting transition temperature is seen only in end-bonded multiwall tubes and the researchers believe that these tubes might contain a less than 1 nm diameter single wall tube at the core. This work gives a great evidence of progress towards one-dimensional superconductors.

The unusual electrical property is actually a manifestation of transport of electrons through single atomic layer that depends on many quantum conditions. Such highly exciting character of CNT demands to acquire an expertise to synthesize specific types of nanotubes in large quantities.

Optical

None of the optical properties of nanotube is found to be highly exciting even though all other properties could be applied to photonics and optics applications. The following would be the important findings from optical measurements:

1. LEDs and photo-detectors made out of single nanotubes show narrow selectivity in wavelength.

2. Optical absorption differs from absorption in conventional 3D materials. Unlike 3D materials, characteristic sharp peaks originating from electronic transitions are observed without any particular threshold. Optical absorption is utilized for first hand judgment of quality of nanotube powders almost routinely. SWNT has absorption edge in near infrared (NIR) wavelength.

3. Photoluminescence is polarized along tube axis and is used to deduce (m,n) indices however, it is inefficient for commercial applications. Photo-luminescence is not observed in multiwall nanotubes.

4. Raman spectroscopy shows high sensitivity and spatial resolution for single tubes and is therefore most popular as the characterization tool showing the graphitic (mode) line typically near 1500 cm^{-1} corresponding to the planer vibrations of carbon atoms as a feature common to most graphitic (like) materials. Raman scattering is essentially resonant in nature which means only those tubes whose band gap is similar to that of the energy of laser light, gets excited. From the intensity ratio of Stokes/anti-Stokes lines, the band gap of individual tubes could be estimated.

This is not an exhaustive list, which as a matter of fact, augments on a daily basis. Nevertheless, some remarkable achievements in applications of these properties have already made breakthrough. It should be mandatory therefore to update another list following the former.

1.4 APPLICATIONS

1. Most noteworthy recent application of mechanical property is the use of bundle of nanotubes to fabricate artificial muscle that have been a long standing demand in polymer science because of an extensive range of utility. The construction of such a thing using exclusively carbon nanotubes is fundamentally new. These are basically nanotube actuators developed from aero-gel sheets drawn from forests of MWNT and they function like muscles under applied voltages. These muscles are capable of operating at extreme low temperatures and no other type of artificial muscles can withstand such low temperatures without getting frozen. Also, the nanotube muscle has effective sustainability at extreme high temperatures where all other types decompose. They work through temperatures of liquid nitrogen to melting point of iron. This new class of artificial muscle could be used to move joints, arms and other structural components in robots, meant for space and planetary explorations where the conditions are unknown and extreme. However, they are yet to qualify as tissue replacements because of the fact that these muscles are to be applied with high voltage for actuation making them operationally incompatible with the conditions of living beings. They sustain through a weight 1000 time more than what is possible using any other type of artificial muscle made out of various other materials. Moreover, the yarns of nanotubes that build the muscle can rotate up to 250 degrees per millimeter of muscle length! This gives the possibility to fabricate miniature rotary motors which are conventionally complex to achieve. Compared to muscles made out of organic conducting-polymer, ferroelectrics and shape-memory alloys etc., nanotube muscles are inexpensive to fabricate and are more useful in making very small valves, stirrers, delivery pumps for medical use, etc. Research is going on to device propulsion systems using these muscles for nano-robots that would deliver drugs within the living systems and remove parasites from the body of living systems. The research outcome of

this joint effort between institutes located in USA, Canada, Australia and Korea is highly appreciatory as an example of coordinated effort in nanotechnology research.

Principle inventor: Ray H. Baughman, University of Texas, Dallas, USA

---

2. Very recently, a hypersensitive pressure sensor has been devised out of a transparent film made of single wall carbon nanotubes. This is another extreme use of mechanical property that a film can tolerate through a very wide range of pressure as low as the pinch between your thumb and forefinger to as enormous as twice that an elephant can impart by standing on one leg! In fact, this transparent sheet is somewhat like a super skin that it can be stretched twice that of its own size in any direction but would return to its normal size and shape once let go. This is made by spraying single wall nanotubes in liquid suspension, on a thin layer of silicone, which is then stretched. The tubes are airbrushed onto silicone in which they orient randomly forming tiny clumps. As the silicone is stretched, some nano-bundles get pulled into alignment along the stretching direction. As the silicone is released, it rebounds back but the nano-bundles buckle up forming tiny nanostructures typically looking like springs. The process is repeated thereafter by stretching the silicone in an orthogonal direction to that of the former stretching. After this sheet is prepared, it can be stretched again and again in any direction without any permanent change in shape! This nano-spring loaded transparent layer is then used to make the hypersensitive pressure sensor in which two such layers are put facing each other on the nano-spring sides by one layer of easily deformed silicone in between. Actually, the middle layer of silicone stores electrical charge like a battery. As pressure is exerted, the amount of electrical charge alters due to the compression in the middle layer. The change in electrical energy (emf) is detected and calibrated to get a measure of the pressure. However, it is not the best sensor in terms of pressure sensitivity but the best advantage in it is that it is transparent i.e. allows light to pass through giving a scope of very unique use perhaps not being hitherto possible!

Inventor: Zhenan Bao, Stanford University, USA.

---

3. Another good application of electro-mechanical property of nanotube is the fabrication of world's tiniest motor that can also be called a nano-motor. The device is made using a multiwall nanotube and about 500 nanometers across of which the gold rotor is between 100 to 300 nanometers. This rotor is attached to a nanotube shaft just a few atoms across and 5-10 nanometers thick. The shaft of multiwall nanotube is welded both to rotor and anchors that are fixed. The complete arrangement can move through about 20 degrees. The real breakthrough in this design actually came from the fundamental discovery of the group a few years ago in which they devised a method to peel off shells from multiwall tubes and grab the core electro-mechanically under transmission electron microscope (TEM). The same technique has been utilized in making the motor also. The outer wall of the nested multiwall tube was broken allowing the rotor and the outer tube to spin freely around the inner tubes, almost like a nearly frictionless bearing! All of the gold

rotor, anchors and opposing stators are simultaneously patterned around chosen multiwall tube employing electron beam lithography. The silicon is then selectively etched creating a trough beneath the rotor to have sufficient clearance for rotation. With applied AC voltage the rotor moves back and forth like a torsion oscillator and the maximum oscillation could go up to gigahertz frequency. This miniature device is surely going to be improved in near future for possible practical uses. Its inventor Prof. Alex Zettl says that gravity has no role whatsoever and inertial effects are basically nonexistent in nano-world. Residual electric fields could play a dominant role here.

Inventor: Alex Zettl, University of Berkeley, USA

4. A very innovative application of the optical properties in cancer treatment has been published in 2009. This joint effort of researchers involving multiple organizations has taken the advantage of two unique optical characteristics of nanotube to detect and destroy the breast cancer cells. These two optical properties are, (1) very strong Raman signal and (2) very strong near infrared absorbance. In a path breaking study, carboxylated SWNT are conjugated with anti-HER2 chicken IgY antibody constructing a unique covalent complex. Raman signal was recorded using 785 nm laser light that detects the conjugate complex within SK-BR-3 breast cancer cells which are treated *in-vitro* with the complex. The Raman signals is found significantly greater compared to that from various control cells. In a second step, irradiation for about 2 minutes with the 808 nm NIR laser @ 5 Watt/cm^2 power could selectively destroy the infected cells without harming the receptor free cells. The SWNTs are not required to be internalized by the cells to accomplish the selective photo-thermal ablation which is indeed a new achievement and the most important advantage of utilization of this complex that might possibly be extended to other types of cancer cells. However, the findings are to be tested through in-vivo experiments and realized through clinical trials to be implemented for medical benefits.

5. Besides these specific novel utilizations, there are many more applications of carbon nanotubes e.g.,

* Field emission displays (FED),
* Touch-screens and flexible displays,
* Filed-effect transistors (CNFET) that are capable of single electron switching,
* As connecting wires between quantum dots in nanoscale integrated circuits,
* As paper batteries where paper thin sheet of cellulose is infused with aligned nanotubes,
* Solar cells formed by a mixture of C_{60} and nanotubes,
* As ultracapacitors,
* As sensors for hydrogen, CO and other unwanted gases, in DNA detection,
* As transparent electrically conducting film to replace indium tin oxide,
* Flywheels for electricity storage,
* As contrast agents in in-vivo MRI,

* For thermo-acoustic sound generation,

* As well-defined tips of scanning probe microscopes,

* For thermal management of electronic circuits, etc.,

* In making lightweight but robust composite materials for next generation aircraft with shape adjustable wings (NASA project),

* As void filler in concrete to prevent cracking due to water accumulation within voids,

* In making lightweight windmill blades by mixing with specific epoxy,

* As electrodes in thermo-cells to generate electricity from waste heat,

* As bacteria detectors in water where antibodies sensitive to particular water-borne bacteria are conjugated with nanotube to form a complex that is deposited onto a paper strip. On contact the bacteria attaches with the antibody and alters the spacing between the nanotubes and consequently the resistance of the assembly is changed.

There are various dossiers available on all of these different applications. Most of the activities are still at research stage but signify the immense potential of carbon nanotube in shaping the technology domination in our daily life in near future. This tremendous use of one material made of a single element tempts to phrase that if *diamond* is the word to signify precious, perhaps carbon nanotube should be referred as the *modern diamond*.

Except the FED panels, none has actually become a truly commercial product as yet because of problems either in the fundamental or technological aspects that are still to be overcome. In fact, there have been a lot of investments till now to utilize the field emission property of nanotubes in designing superior TV screens by many entrepreneurs since its discovery. However, a major success is still about to come. It could be though relevant to enumerate the development in order to have a guess on the typical time scale of the journey from lab to market that any application is subjected for a commercial success!

Two striking specialty of carbon nanotube are, (1) it has a very high aspect ratio that man could have possibly synthesized for any material and (2) highest strength to weight ratio for any known material. Apart from these two natural attributes, the other most important feature is the emission of electrons from nanotubes under the influence of an electric field, the field emission. This property was discovered very quickly following the discovery of nanotubes. Very soon its importance has been realized as another potential alternative to replace the cathode-ray tubes (CRT), the household item that used to be the picture tube of the first generation TV sets and computer monitors. With the advent of liquid crystals, flat panel displays (LCD) has started dominating every segment of the display market.

For large display panels of course, the industry also has invented the plasma technology and several varieties of rare projection technologies, latest even using the light emitting diodes (LED) etc. that is now capturing the market worldwide. The primary reason of fast adoption to flat screen display is weight and size reductions, i.e. an improvement in the *form-factor*. However, (1) power consumption rate has not reduced drastically and (2) the overall image quality is not better in the rival flat panel displays available at market today compared to that in the CRT displays. Nanotube and its field emission property usher the promise of combining the image performance of CRT with the form factors of flat panel displays. Technologically, this is not providing something new but certainly bringing in an advantage in terms of quality! It would harness the same light generating mechanism that of a CRT, however in a very small space and with much higher resolution because being small, the physical number density is higher and since each tube is a source of electrons, hundreds of them are placed directly behind each pixel. CRT has only one electron source that can write to one pixel at a time. Therefore, the resolution and effectiveness in motion rendition enhances multi fold when nanotubes are put for displays and the very fact that the electron sources are just a millimeter away from the phosphor screen, by default due to their tiny physical size, the displays are ultra thin which is incomparable to anything envisioned in such product line! Other types of field emitters have also been tried using micro lithography technique. But they did not turn out to be as good an electron emitter as the carbon nanotube. Also, because of the covalent bond, the structure is incredibly stable here under high electric fields and because of large aspect ratio the image essentially becomes extraordinarily sharp. There is no additional problem because nanotubes do not adsorb any harmful gases like the conventional materials. Compared to lithographic products, nanotubes are inexpensive to make and fabricate a panel using self-assembly. Sony, Samsung, Motorola, CDreams etc. companies have already produced their prototypes that could be taken as a proof of the concept. Nevertheless, there is still time that it might be expected to be on shelves. If discovery has the pain of giving birth, bringing it into the stage of a product at market has the ache of nurturing the baby to make him stand on his feet! As a matter of fact, it is our experience that the facts and figures accelerate at every home as baby grows through the events of passing days.

2. Titania in Nanotechnology

Titanium dioxide (TiO_2) or Titania is naturally occurring and is commonly referred as Pigment White 6. It is the most widely used pigment, having an annual global consumption of four million tons. Also, besides large scale utilization as pigment, TiO_2 has several other common utilities, e.g. natural whitener, permanent markers, UV absorbers, photocatalyst etc. Some of the applications of TiO_2 are:

* In cosmetics as a UV absorber, which is already in commercial use.
* In making very large capacity (~ 25 terabyte latest) media for electronic data storage.
* In protein cleaving at the amino acid Proline site. This is useful in Proteomics studies.
* As a photocatalyst, it is being used in many new applications ranging from environment to energy. The photocatalytic property was discovered by Akira Fujishima way back in 1967.
* As nanoparticles in creating self-powered electronic devices and displays.
* In hydrogen production as fuel.
* Waste-water treatment.
* Food packaging.

2.1 ENVIRONMENTAL APPLICATIONS

Our indispensible dependence in modern life on machinery has given us huge advantage and comfort but at a cost of contaminating the environment. Nevertheless, a cleaner environment is a global mandate! Cleaner air is the major part of a cleaner environment. Air pollution has become a major challenge around the globe today, which is very difficult to do away with; moreover, it has the effect beyond any geopolitical boundary. It is unfortunate that we still burn materials for generating electricity and transportion. A complete environment friendly solution meeting all our requirements is indeed difficult to materialize. Along with our search for potential alternatives, one other measure is to look for materials and processes that can help in continuous small scale cleaning, if not an automatic one like photosynthesis which happens to be the only natural process to clean up the air. In fact, the extent and nature of pollution in today's world is complicatedly much larger in both size and character that is hardly manageable by only doing additional planting. In fact, the complex nature of pollution in modern world has also jeopardized the survival of the botanical world. Therefore, it is extremely important to look for new materials and processes that will be able to assist the process of cleaning up the atmosphere preferably, concurrently. Titanium dioxide could be a potential achiever in this regard.

Breathing cement

Imagine that all buildings in a city can be made to be the silent workers by trick of the trade for the job of cleaning up the air! Titania can make this happen. Photocatalyzer TiO_2 can eat up the smog from your atmosphere leaving you with the clean air. Along with improved fuel and strict emission control, it is possible to really have clean air even in cities, automatically for all time, if the city structures are made to be engaged in the business of cleaning. It is not a fairy tale but has been achieved! Itlacementi, the largest Italian cement company has claimed to have achieved it using TiO_2 and legitimately bagged the 'Times Best Invention Award' in 2008 for their product *TX Active*. Nanophase TiO_2 incorporated into cement, mortar, paint and plaster engages in breaking down the exposed pollution in air from industrial activities, car emissions, smoke from heating systems etc. when light shines on these new hybrid materials. The photocatalytic action helps reducing the pollutants in air directly. Measurement shows 45-60% reduction in nitric oxide contents in an atmosphere over an area of two acres. That indeed is quite a remarkable result for the application of a first of its kind product, which definitely raised a hope for more improved varieties through process and product research. Therefore, building tall new city structures and covering the existing ones with this new class of building materials, certainly makes sense in using it as well as engaging it in cleaning at the same time. Quite obviously, the global market potential is immense for any improved process or product in this regard.

Self-Cleaning

The photocatalytic property of TiO_2 and its nanocrystalline form have wonderful utility in making glass or specific plastic as self-cleaning. Usual practice of making such a glass is to use hydrophobic coats. However, there are several drawbacks in any or all of the hydrophobic coatings that prevented their widespread use. Whereas, TiO_2 achieves the same effect through quite contrary phenomenon and instead uses the hydrophilic property that also washes off the dirt form the surface. When light falls on a thin titania-coated glass, photocatalysis causes the chemical breakdown of adsorbed organic dirt. The hydrophilicity of the process creates a thin sheet of water on glass that washes off the dirt. There is also a conversion of relevant Ti^{4+} sites to Ti^{3+} sites that favors the dissociative water adsorption. Among all polymorphic structures, the anatase phase titania is found to be the most effective for this process. There are several ways to form the thin coatings of nanocrystalline TiO_2 at lab scale that are useful to deposit 10-30 nm layers. The coated layers could be made robust through specific post-treatments. The advantage of using TiO_2 in such commercial production is that it is non-toxic, chemically inert in the absence of light, further to being inexpensive and already a familiar household chemical known through many common applications. In today's modern glass architectures in advanced cities around the globe, use of such glass would be indispensible to keep them ever shining. When it removes the dirt deposited from atmosphere in a very large scale through huge areas of exposed glass, it definitely also contributes in making the air around cleaner.

Anti-fogging

Commonly, surfactants that minimize the surface tension of water is used for anti-fogging applications, e.g. soap, detergent, shampoo, shaving cream, gelatin, hydrogels and some colloids and nanoparticles. Essentially, it is the wetting phenomenon that helps creating a non-scattering film of water instead of droplet formation, as droplets are translucent that reduces the visibility. The hydrophilicity under exposure of light in titania makes it an excellent agent in making anti-fogging glass or plastics. Imagine, its utility in many important sectors and worldwide market potential of having a cost effective development for use in daily life to outer space exploration!

2.2 RENEWABLE ENERGY APPLICATIONS

Nineteen years after Fujishima's discovery of photocatalysis using TiO_2, another landmark discovery has been made using the same material by Brian O'Regan and Michael Grätzel at École Polytechnique Fédérale de Laussane (EPFL), Switzerland. Their genius has been to utilize the thin film of TiO_2 nanoparticles in designing a solar cell and successfully generate sizeable current density that people could not achieve by utilizing many other materials. Thus it became a new class of solar cells, referred as *Grätzel Cell*. The fundamentals of their invention are published first time in 1991 and Grätzel was honored with the *Millennium Technology Prize* in 2010. This device is basically a photoelectrochemical system and because of the fact that it uses a dye that sensitizes the semiconducting TiO_2 film, it is also called *Dye Sensitized Solar Cell* or DSSC. Use of the semiconducting film made out of nanometer-size TiO_2 particles has increased the effective surface area and roughness that could adsorb much more number of the charge-transfer dye molecules, resulting into a major 46% harvesting of the incident light. In fact, electron is released here from the dye molecule and injected into the conduction band of TiO_2 and this mechanism itself is fundamentally different from that in a traditional semiconducting cell, giving an advantage of continuous operation even at high temperatures. Thus, an alternate mechanism for solar power where charge carrier generation and their transport happens in two different materials has been initiated. Instead of describing the fundamentals of underlying physics of its operation, it would be pertinent here to have a comparative account of DSSC with respect to the traditional cells:

* DSSC costs less compared to cells made out of elemental semiconductors and thin film technology.
* They are easy to construct and do not involve any sophisticated fabrication procedure. In fact, hobby kits are available to hand construct a test DSSC.
* Traditional silicon cells are fragile and needs to be properly caged with metal backing for strength. There is a decrease in efficiency as cells heat up internally. Whereas DSSC are built with only thin layer of plastic on the front, allowing them to radiate the heat away much faster and therefore more suitable for continuous operation.
* The very fundamental change in the trick of the trade where functions of electron generation due to incident light and the charge carrier transport are separated, has effectively enhanced the stability over 5 million turnovers without decomposition. Continuous operation for long period over months together is possible and the cells are durable at much high temperature. There is hardly any loss of efficiency. In traditional cells, the two functions being happening within the same material, unwanted situation occur giving rise to loss in efficiency.
* The monolayer coating of the charge-transfer dye in DSSC essentially shifts the absorption onset to 750 nm and the light harvesting efficiency reaches almost 100% for all visible range. This makes it suitable to work even in low light conditions, e.g. under cloudy skies and non-direct sunlight or even indoor lights. Overall conversion yield even in diffused light is found to be 12%. Traditional cells suffer a cut-out at some lower limit of illumination when charge carrier mobility is low and recombination becomes a major issue.
* Because of mechanical robustness and light weight glass-less collector in DSSC, it is highly recommended for low density application like rooftop solar collectors. In fact Dyesol and Tata Steel partnership have developed the largest such photovoltaic module printed onto steel in a continuous line in June 2011. This is a groundbreaking achievement in commercialization of DSSC technology.

* The overall light harvesting capacity of dye cells could be made up to about 50%, however the current is still limited to 20 mA/cm^2 unlike 35 mA/cm^2 in silicon based solar cells. Power production efficiency is about 11% that has made it less competitive in commercialization.

* The major disadvantage of DSSC is the use of liquid electrolyte that has problem at low temperatures where freezing stops power production and even causes physical damage to the cells. Very high temperature too can cause possible leakage. Gel electrolyte molecules have been created that is transparent and non-corrosive. Further, research on performance of new state-of-the-art devices in this class is continuing worldwide as the scopes are plenty and above all, the winner would have the opportunity of the major share in global market!

The other TiO$_2$ product in this regard is the very recent discovery of a *Solar Paint,* using the *quantum dot* technology. These quantum dot solar cells contain a binder-free paste made of cadmium sulphide (CdS) or cadmium selenide (CdSe) and TiO$_2$ nanoparticles applied as paint on the surface of a conducting glass and annealed to 200 degrees centigrade temperature. One percent power conversion efficiency is achieved using conventional paintbrush approach under ambient conditions. This is however five times less than the highest recorded efficiency in quantum dot semiconductor (QDSC) prepared by multi film architecture of mesoscopic TiO$_2$ layers followed by layering of CdS and CdSe. However, constructing an anode with simple brushing of paint is much faster than making it by multi film approach.

2.3. TITANIA NANOTUBE

The anatase TiO$_2$ can be converted to produce tubular nanostructure through hydrothermal synthesis. These are called Titania Nanotube (TNT) typically having 6-12 nanometer inner and outer diameters with a maximum length in the range of 1 micron. A self-assembled array of TiO$_2$ nanotubes can also be grown through anodization in electrochemical cells directly on top of a titanium metal surface. TiO$_2$ nanotube is a unique new material and different from carbon nanotube (CNT) because of its molecular bonding. Such nanostructures have huge surface area compared to their size and are already finding several important applications in solar and battery technologies, besides the nanaotechnology for food. For example, recently, the application possibilities of this material have been demonstrated to devise hybrid novel anodes for a new high-performance battery and as sensors to certain bacteria that contaminate food.

2.4. Self-Powered Display

A new TiO_2 based product is on the horizon that has the potential to create a boom in the cellphone market. The displays will not only show images but also generate power to charge the phone when not in use. Nokia has developed a 200x200 pixel monochrome prototype. TiO_2 nanoparticles are used here to generate both power from light and image from applied power. The screen runs on forward bias when light shines on it to generate electricity that charges the battery whereas in reverse mode, it flashes a black image when power goes into it using the control electronics that operates as the buttons are pressed. There is a huge possibility that it would initiate a major change in cell phone technology in the near future. It may be worth remembering in this context that Market helps those who help themselves!

2.5. Polymer and Textile

The application of nanotechnology in polymer research to obtain advanced textiles that are stain repellent, wrinkle free and have no static charge effect has become an active area. Leading international fashion houses like Hugo Boss, René Lezard, Paul Stuart etc. are spending considerable sums and effort in the research for such products that have huge target market. Camouflaged clothing that can change color with the intensity of daylight and specific garments that can be engraved with nanoscale sensors for continuous remote monitoring of the physical parameters of a human body under critical ailment are demanding on the development scale. TiO_2 has a role here as well. Fabric treated with TiO_2 nanoparticles can replace the fabrics with active carbon that is used commonly as chemical and biological protection fibers. The photocatalytic activity of TiO_2 can break the harmful and toxic chemical and biological agents. These nanoparticles can be pre-engineered to adhere to the textile polymer with advanced spray coatings or electrostatic deposition. Their pressure sensitivity is also useful for making sensor-incorporated fibers. Research is being carried out to integrate piezoceramic nanocrystalline particles with TiO_2 nanoparticles to produce garments that might work as a clinical device for constant monitoring of the body parameters of critical patients. Imagine the cost effectiveness in terms of replacement of many voluminous and cumbersome instruments at ICUs of hospitals!

2.6. High Density Data Discs

The Japanese research team at Tokyo University has developed a new material based on 5-20 nm size titanium oxide nanoparticles that can be used to make data storage disc having capacity 1000 times more than a blu-ray DVD. It is in fact a new nanocrystalline form that shows a reversible photoinduced metal-semiconductor phase transition between λ-Ti_3O_5 and β-Ti_3O_5 at room temperature so that it transforms from a metallic black to brown semiconducting state when light shines on it. This unusual type of light dependent electron transport characteristics makes it a very promising material for next-generation optical storage device because λ-Ti_3O_5 nanocrystal is a good candidate for optical storage using near-field light and the memory density could be as high as 1 terabit per square inch. It might also be possible to store different types of data in different color codes using such a storage media. Commercial production of such devices is underway through private sector participation with academia.

A reverse solar cell having a proton exchange Nafion membrane with TiO_2 photoanode, platinum cathode, and acidified methanol as the electrolyte have been successful to continuously generate hydrogen under UV excitation. Wastewater treatment using TiO_2 on the other hand, has continued over the last two decades. Photocatalytic degradation has been applied to remove pollutants like aliphatic and aromatic organics, dyes, surfactants, several insecticides, herbicides, and pesticides etc. from waste-water by converting them into CO_2, H_2O, and mineral acids. Acid-blue 25, as a waste from the textile industry is now routinely removed using TiO_2 suspension in pilot scale.

In the history of scientific and technological development, many a time it is found that the familiar things often go unnoticed, unattended and later discovered to be actually not so well known. Titania is one such material existing in our civilization since its discovery in 1791 from ilmenite ore. Our understanding in modern science and technological capability to make and maneuver materials at nanometer scale could reveal this old and familiar material's many hidden potentials for our benefit. Realization of these immense possibilities should be an important objective. That would mean another impetus in ushering a new acceleration in society. After all, *Technology* is that unstoppable force that dominatingly drives our civilization forward and *Market* is the exchange that we sustain for paving the road for technology. TiO_2 might become the gold if carbon is recognized as the diamond for nanotechnology!

3. Nanotechnology in Textile, the NanoTex

Imagine you are at a party and a little careless that the red wine spills out of your glass onto your costly beige-coloured suit! But you don't worry because your suit cloth is such a material that it does not wet and nothing on earth sticks to it and liquids just roll down! No proof of an incident! You are back in a good mood and proud to be the owner of such a suit, on top of the world! Or, imagine some-one close to you is sick and needs continuous medical attention and the body parameters have to be monitored throughout 24 hours. You expect hospitalization under intensive care where each patient is connected to several monitoring machines with wires and cables attached or fastened to the body, i.e. immobile even though able to move! But, what if these are not there! What if the patient wears special clothing that gives signals for individual body parameters that can be addressed and recorded under remote control? The patient is free to move around but the blood pressure level is monitored continuously!

All these and many more are near possibilities as the technology of textile is mar-ried to that of nano, i.e. the new textile, which Burlington Industries subsidiary named 'NanoTex.' Their products are claimed to resist spills and static charge, repel and release stains and provide the careless comfort. This is however a story with only garments and apparel that are actually a part of the larger research activity of polymer in nanotechnology or nanotechnology with polymers! There are not only the possi-bilities of textile with advanced properties as mentioned above but also smart textiles with properties that could never have been imagined for textile so far. In fact, in nan-otechnology, the first commercial success has been achieved in textile!

A distinction could be made looking into different kinds of manufacturing that includes nanoscience by way of whether it involves the synthetic nanoparti-cles or fibers of nanometer dimension. Even though such classification cannot really be strictly maintained because of overlaps on several material aspects, it could be convenient though for the sake of having a consolidated description about various activities in different directions.

* *Treated Textile*: These are known textile materials treated with the deriv-atives of nanotechnology (i.e. nanoparticles, materials, coatings etc.) to realize certain new or functional material or introduce a novel aspect into the properties of a known fabric. Nanoparticles can be introduced into the fiber or synthetic material to spin out the fiber, resulting into a nano-com-posite. The motive here is to improve the existing functionality.

* *Smart Textile*: These are products obtained through new technology for textile, derived from concepts of nanoscience. The motive here is to real-ize new materials with functional applications. For example, nanoscale fibers can be made from cellulose utilizing improved techniques and machinery that would produce a new quality textile. The objective here is to have a new property or combinations of functionalities.

The salient features of these two categories could be summarized as follows:

Treated Textile	Features
	1. Water, oil and dirt repellent
	2. Stain, spill and static resistant
	3. Wrinkle resistive
	4. Anti adhesive
	5. Reduced need of washing
	6. Durable without altering the feel of fabric
	7. Highly moisture absorbing (polyester)
Smart Textile	Features
	1. Chemical and toxic resistant
	2. Odor controlling
	3. Anti-bacterial
	4. Fire resistant
	5. UV protection
	6. Self-cleaning abilities, e.g. smart surgical gloves
	7. With integrated electronics as nano-sensors
	8. Form processed nonwoven fabric
	9. Conductive fibers
	10. Conductive pressure-sensing fabric
	11. Luminescent polyester

3.1 METHODS OF FABRICATION

3.1.1. Water, Oil and Dirt Repelling

3.1.1.1. NANO-WHISKERS

Nano-whiskers are hydrocarbons 1,000 times smaller in size than a typical cotton fiber. A kind of peach fuzz effect appears when these whiskers are mixed to cotton fiber. It does not reduce the strength of the fiber. The space between the whiskers in the fabric is less than the typical size of a visible water droplet which therefore cannot penetrate through and resides on top of the fabric surface and eventually drops out. However, the spacing is still larger than the size of a water molecule and therefore, water molecules would pass through if pressure is applied. The fabric can still become wet and breathe to give comfort and is not like plastics! Burlington Industries subsidiary NanoTex has applied this method in their fabrics.

3.1.1.2. NANO-SPHERES

Another prominent method is extracted from the biomimicry of the *lotus effect* in *nature*! The Swiss based textile company Schoeller has followed this by developing *nano-spheres* that are saturated into the fabric. This involves a three dimensional surface structure with gel-forming additives that repels water and prevents dirt particles to attach. This is similar to what happens in lotus petals. Water beads up and rolls down due to gravity taking also the dirt along its way!

3.1.1.3. PLASMA-COATING

The hydrophobic or repellent property could be achieved as well by coating the fabric with nanoparticle plasma film. This method also gives super-hydrophobicity without losing the softness of the fabric. The best example in this product is the swimsuit that world champion American swimmer Michael Phelps used at Beijing Olympic. This fabric has plasma layer coat of nanoparticles to repel water molecules, thereby reducing the frictional pressure enormously as compared to common swimsuits. Though costly, this material has become popular in swimming competitions nowadays.

3.3.2. Nanoparticle Coatings

Plasma coating technique is used to produce fibers, which have excellent hydrophobic as well as oleo-phobic characteristics. In fact both plasma-assisted and conventional deposition coatings could be used to coat with nanoscale siloxane layers that bond covalently to the fibers and exhibits super-hydrophobicity. For stain, dirt and oil repelling characteristics, fluorocarbon impregnation has been shown to exhibit better result and it does not seem to affect the other qualities of textile.

In fact, coating of cotton fabric with different nanoparticles has become a major trend in introducing various new functional properties into the fabric. Semiconductor oxide nanoparticles, e.g. TiO_2, ZnO, SiO_2 and Al_2O_3 have been used to obtain UV blocking, stain-repelling, odor-removing as well as anti-bacterial characteristics. In fact, TiO_2 and ZnO have been found to be useful in achieving a number of novelties. They both have photocatalytic property; the only difference is that the former has a band gap of 3.2 eV whereas it is 3.37 eV in the latter. Nanometric thin metallic layer is found to be useful to have electromagnetic shielding, antistatic, heat dissipation and also UV protection characteristics of fibers. Nano-sized silver (Ag) particles are used to attain anti-bacterial character while platinum (Pt) and palladium (Pd) particles are used to decompose toxic gases, an important application for bio-hazards and possible prevention of biological warfare. Even, the organic coating, e.g. polytetraflouro-ethylene (PTFE) has been used to create antistatic fabrics. The textile out of this fabric should prevent the wearer from electric discharge and such fabric should as well be suitable for electricians working on high-voltage supply lines. This organic coating is very stable and retains through several cycles of laundry, which is an important criteria in coated textiles if not meant only for special single use, e.g. in medical applications such as caring for patient's wounds or during surgical procedures.

3.3.2.1. COATINGS WITH PHOTOCATALYTIC NANOPARTICLES

Titanium dioxide (TiO_2) and Zinc oxide (ZnO) are the primary materials in this category, which are used as coating to block UV, odor-eating and dirt-removing actions. The photocatalytic reaction is not any different from that of its utilization in making the DSSC solar cells described in previous section. The reaction is shown in schematics in Fig. 12. When these materials are illuminated with light in which the photon energy is higher than the band gap between the conduction and the valence band of these semiconductors, the electrons (e^-) jump out of the valence band into the conduction band. In the process a hole (h^+) is created in the valence band. This (e^-—h^+) pair is on the surface of the photocatalyst material. Electrons combine with oxygen in air and create $\cdot O_2^-$ whereas the h^+ create the hydroxyl radical $\cdot OH$ from water. Since both of these are highly unstable and reactive, when it comes into the contact of an organic compound, CO_2 and H_2O are produced and this oxidation-reduction reaction continues in cascade. Thus, the odor-creating organic molecules are broken continuously and thereby the odor vanishes. As UV light normally has energy greater than the band gap, such a coating thence gives also a shielding effect from UV for the users of the textile made out of these materials. The water molecules generated in the process also drains out the dirt which is wedged deeper within the unevenness inside the fabric. The textile made out of these coated fabrics is thus capable to protect the wearer from UV while auto-cleaning and removing bad body odor at the same time. It has also been determined that fabric treated with nano-TiO_2 could provide protection against discoloration of stains. Moreover, because of these properties, such textiles almost do not need any laundry cycle. If at all necessary, ultrasonic cleaning is recommended as better than the detergent cycling.

3.3.2.2. ANTI-MICROBIAL COATING

Nanoparticles of silver (Ag) metal are found to be most effective in this category even though both TiO_2 and ZnO could be used as well. This is because metallic ions and compounds display certain degree of sterilizing effect and adversely affect the cellular metabolism inhibiting the growth of a cell. Small nano size particles can penetrate through the cell membrane when it comes into direct contact with bacteria and fungus. The smaller the particles are, greater is the total effective surface area of an ensemble of such particles that makes the contact. These particles inhibit the multiplication and growth of the bacteria and fungi that cause infection, odor, itchiness and sores. The textile for making socks is ideal to apply silver nanoparticle coating. It is also useful for clothes used for dressing of burns, scalds and at skin donor or recipient sites etc.

Beijing JLsun High-Tech Co., Ltd. has been the pioneer in such antibacterial textile products. However, the coating recipe is under trade secret and technically referred as SCJ-875 antibacterial finishing agent that have been modified later to SCJ-877 and SCJ-963, which are claimed to be the top quality in the world in this regard. Their products have been certified by the Chinese Academy of Medical Sciences. It has been claimed that the textile treated with agent SCJ-963 has anti-bacteria, anti-odor, anti-inflammation and mildew proofing characters. It relieves itching and has astringent functions. It inhibits many bacteria like, staphylococcus aureus, klebsiella pneumoniae, salmonella typhoisa, escherichia coli, pseudomonas aeuginosa, bacillus subtilis, candida albicans, penicillium nuatum, aspergillus niger etc. and the rate of inhibition has been claimed to be 99.95% even after 100 washes. The textile is non-stimulant, non-allergic and non-toxic to skin exposure. The fabric is also claimed to prevent from a number of infectious diseases and infections from, e.g., ringworm of the foot, eczema, sweat and foot odor, skin itches etc.

3.3.2.3. HEALTH IMPACT OF NANOPARTICLE COATED TEXTILES

The health impact of nanoparticle exposure to bare human skin and body is a very recent subject and not at all developed yet. There are reports and counter reports leading to confusion on the effects of exposure. However, for use in fiber vis á vis textile, it is indeed an important question no less than the question of wash sustainability of textile. Precautions are always taken in synthesizing nanoparticles under laboratory conditions in regard to human body exposure. Since the particle size is smaller than the pore size of our skin, it has been suspected that the particles get into the bloodstream by penetrating the skin and land on a cell. Therefore, the reaction of particular nanoparticles with the biological entities in living organism is extremely important but is a huge area of study that indeed is growing slowly. The caution here is not to become a victim of any new prejudice that might hinder the progress in science! Nevertheless, it might be pertinent to introduce some kind of international certification as a primary label for commercial practice and subsequent use.

Materials that are mostly in use as nanoparticles in textile are TiO_2, ZnO, Ag, Al_2O_3 and SiO_2 etc. Yet the effects of natural skin flora have not been tested or reported conclusively. There are many sample studies as such on the carcinogenic effect of TiO_2, which has been proved to be safe for use. However, the effects of TiO_2 nanoparticles are yet to be evaluated. As mentioned in previous paragraph, nano-silver is used for clothing, which is supposed to protect people suffering from neurodermatitis, an infection exacerbated due to the bacteria staphylococcus aureus. However, clinical studies across the world have not yet confirmed the use of bare nano-silver coated textile. The German Federal Institute of Risk Assessment (BfR) has not recognized the anti-bacterial effect of such a textile and warned against the negative side effects such as a weakening of the immune system and possibility of developing silver resistant bacterial strains, etc. It also cautioned against a false sense of security and neglect of general hygiene.

3.1.3. Nanofibers and Embedding

A variety of nanometer diameter polymeric fibers, e.g. from PA, PEO, PLA, DNA, starch etc. have been produced using an electrostatic spinning method. Polymer solution is sprayed into a high electrostatic field. Fibers are produced because of electrostatic charging and surface tension of the solution. A thin polymer jet is ejected in spinning as the surface tension is overcome by the electric force on induced charges. Long strands of fibers having diameter in the range of 100-200 nm could be produced in this process. Nanofibers have also found application in making sophisticated blood filtration systems for clinical use.

Attaching 20-50 nm thick polymer nanolayers to natural cotton fibers, chemical-resistant textiles have been fabricated with the brilliant property of selective transport as found by researchers in North Carolina State University and the University of Puerto Rico. The cotton fibers are coated with layer-by-layer (LbL) deposition of nanometer thick PSS (poly sodium 4-styrene sulfonate) and PAH (poly allylamine hydrochloride). The nanolayers adhere to fibers by electrostatic force. The coated fibers could be customized for different chemicals so that it is possible to specifically block warfare agents like mustard and nerve gas or industrially hazardous chemicals, however allowing air and moisture to pass through, keeping the fabric breathable. Chosen chemicals bind to the polymers of the fibers and are blocked from passing through while the gaps in the textile can easily pass air and moisture. There are many practical applications of these fibers, e.g. diapers coated with anti-itching polyelectrolyte, tissues coated with anti-allergy medicine etc.

Mono-dispersed particles, comparatively larger in size prepared typically in wet-chemical and aerosol method, are embedded within a polymeric matrix which could be transformed into fibers. Particles of few tens of nanometers of metals, metal-oxides, metal-salts and polymeric substances are introduced to realize new functionalities in natural fibers and improve the fiber properties. It is anticipated to produce more wash-proof textile.

Composite fibers reinforced with clay nanoparticles are found to exhibit flame retardant and anti-corrosive behaviors. The most commonly used, clay *montmorillonite*, has been applied as a UV blocker in nylon composite fiber. Dense packing of nanosize clay flakes introduces barrier performance to water, chemicals and other harmful species. The mechanical properties of the fiber improve, e.g. the tensile strength is found to increase by 40%. Further, the heat distortion temperature also increases by a margin of about 80°C.

3.1.4. Smart Tex

This category of fabrics is the one that has the primary objective of imparting new functionalities in textile. For example, textile with integrated electronics, special self-cleaning abilities, resistance to fire, protection from ultraviolet light etc. and a range of other innovative features. These have applications in aerospace, automotive, construction, electronics, healthcare and sports. Self-cleaning surfaces, smart surgical gloves, implants, prosthetics, clothes that can sustain extreme environment, e.g. extreme hot or cold or toxic conditions and garments with sensors that monitors round-the-clock the body parameters of a sick person etc. are a few things anticipated in this class of textile material.

Toray Industries Inc. has developed a nano-scale processing technology called *NanoMATRIX* that forms functional material coating consisting of nanoscale molecular assembly on each nano-filament that forms the fabric. This technology aims to build new complex functionality without losing the texture of the fiber. The concept of self-organization is applied in this technology. Precise control of physical conditions like, pressure, temperature, magnetic or electric field, humidity, additives etc. associated with the interaction between the functional material and the fabric polymer, helps to create the coatings in the range of 10-30 nm thickness while the molecular assembly of the functional material is achieved with high degree of uniformity which could not be realized in other types of synthesis techniques. This is actually a *bottom-up* approach of synthesis in nano-textile and solves many after treatment issues. Due to the extraordinary level of precision in uniformity in the polymer coating, it becomes possible to improve the functionality and durability to a much higher degree that is difficult to achieve in other methods. In addition, being a slow buildup process, it is claimed that combinations of functionalities having conflicting properties could also be managed within the same fabric, e.g. anti-static function requiring a certain amount of water absorption and water-repellent action keeping water off altogether, might be achieved by controlling the state of molecular arrangement and assembly of functional materials on each monofilament.

SmartShirt developed by Sensatex is another example of smart textile that incorporates microscopic wires interwoven into the fabric itself which work as sensors to monitor the physical parameters of the wearer, such as heart rate, temperature, UV exposure, caloric burn etc. This material is in general referred as *integrated smart textile* that incorporates foreign bodies, tiny devices and/or equipment etc. within their fabric framework. *SmartShirt System*™ is one such product patented and launched by Maryland (USA) based Sensatex. This is a seamless knit shirt integrating a conductive fiber system to wirelessly carry physiological signals from the body. Application research and reliability studies with this system are being taken care of now. Possible use of this system outfit enabling monitoring and data-feeding in real-time could be the home health monitoring system of the elderly, observing outpatients in post-operative and chronic illness situation, training support for athletes, remote monitoring for first responders, workers for hazardous materials, soldiers in the field, and tracking through the vital signs of long distance professional truck drivers to spot alert them for fatigue etc. The system has promise for health care professionals to use it to have an early warning of abnormalities in patients. With the possibility of remote access, the item would be in demand in today's busy world in order to measure and monitor the vital functions of patients stationed miles away from the health care center.

3.2. PRODUCT CERTIFICATION AND QUALITY CONTROL

Already there are quite a number of companies making nano-textile products and the anticipated market is quite large. In this way, textile has been the pioneer in bridging the new frontier of technology from laboratory to home. Woodrow Wilson International Center for Scholars listed a total of 156 items under the *clothing* category on its nano-product database. There are about 82 products reported that could be listed under the same category in European and Asian market. Hohenstein Institute, a private research and service organization in Germany has also introduced a quality label for nanotechnology products. A textile product does not qualify *Hohenstein Quality Label* simply by incorporating nanoparticles within the fiber or performing a nanoscale coating! The product should show the demonstrable result of a new function without any adverse side effect or resistance to care treatment or wear comfort. This is important as criteria; however its implementation requires the support from administrative authorities of a country and also the market. For consumers, such a label is handy in order to differentiate a genuine nano-textile product from that in which *nano* is only a powerful sales catchword. Compared to the number of product information inundating the internet every day, the number of registered ones seems much lower.

The current statistics might have changed but that is not the real issue. Important is to have an international statutory certification and validation authority established under the common consensus or the United Nations that would have to be concerned when it comes to the question of use in healthcare or direct contact to human body. There are indeed lots of unknown effects on human health of nanotech products. Caution should be there but not to the extent of inhibition that hinders the progress of technology. In fact, international trade regulation might control the use of any particular product in any particular region but, the effect on human health, if any, would transmit without caring the law of any particular land and will pervade across all boundaries!

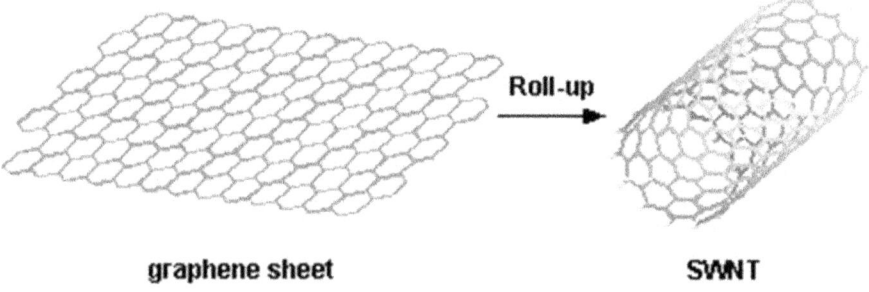

graphene sheet   **SWNT**

Fig. 9. Single graphitic sheet (or Graphene) and the rollover to form the tubular structure call Carbon Nanotube (CNT). Depending on how the sheet rolls over itself, different types of tubes are formed as shown in Fig. 10.

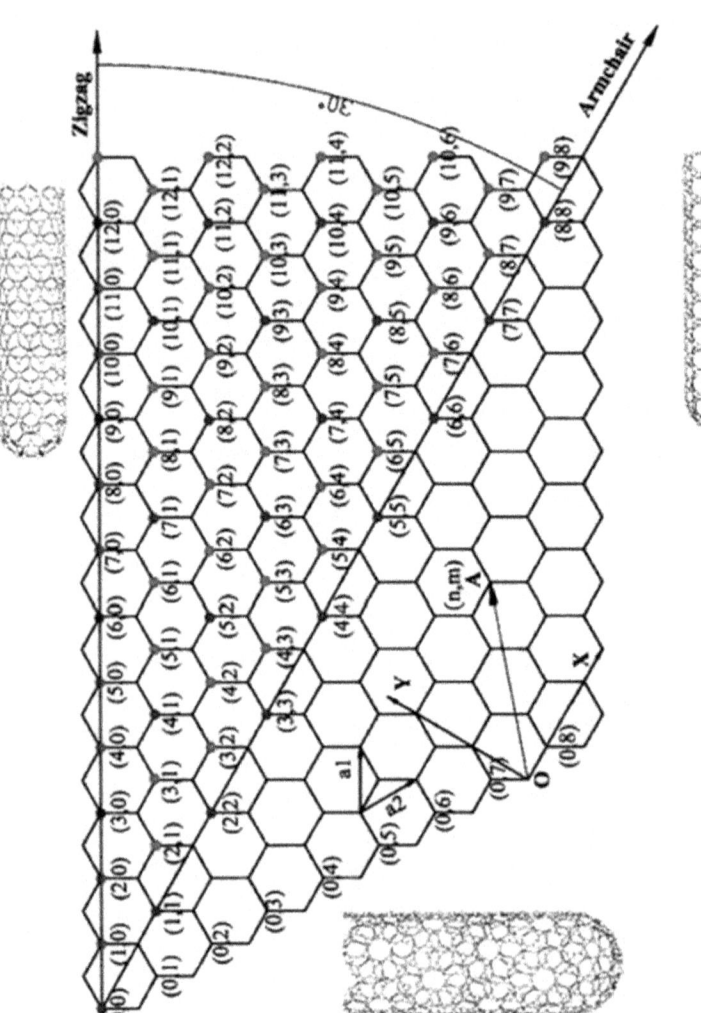

Fig. 10. The formation of 3 different types of CNT has been shown through the rollover of single graphitic sheet in 3 different ways: tube at the top is called Zig-Zag; the bottom one is known as the Armchair tube while the tube shown vertical is the Chiral one.

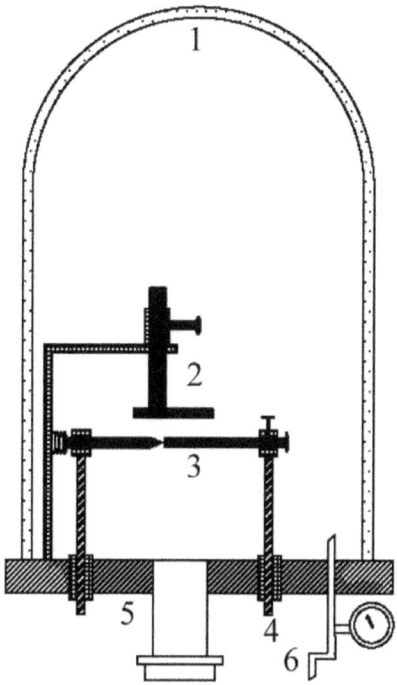

Fig. 11. Schematic drawing of a typical simple Smoke-Source for producing CNT through arc-discharge: 1. Bell-Jar, 2. Smoke catcher, 3. Graphite rods, 4. Electrodes, 5. Vacuum pump, 6. Gas inlet and Pressure gauge.

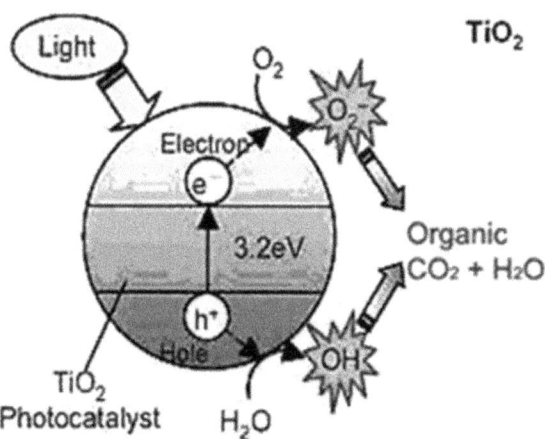

Fig. 12. Photo-catalysis in Titenium Dioxide (TiO_2) is shown in schematics.

Bibliography

Books and Journal Articles

Abrahamson, John, Peter G. Wiles and Brian L. Rhoades, **Structure of carbon fibers found on carbon arc anodes**, *Carbon*, vol. 37, 1999, p.1873 (*original conference paper of 1979 republished*)

Aliev, Ali E. et al., **Giant-Stroke, Superelastic Carbon Nanotube Aerogel Muscles**, *Science*, vol. 323, 2009, p.1575

Alivisatos, P, Whiteside G, *in* **IWGN Workshop Report**, 1999, p.1

Alivisatos, A. P., W. Gu and C. Larabell, **Quantum dots as cellular probes**, *Annu. Rev. Biomed Eng.* Vol. 7, 2005, pp.1235-39

Ardnne, von M. **Improvements in Electron Microscopes**, *GB Patent*, No. 511204, 1937

Atala, Anthony **Regenerative medicine**, TED talk, March 2011

Avouris, Phaedon, Lyo In Whan, **Observations of Quantum Size Effects at Room Temperature on Metal Surfaces with STM**, *Science*, vol. 264, 1994, pp.942-45

Azzazy, H. M. E., Mai M. H. Mansour, **In-vitro diagnostic prospects of nanoparticles**, *Clinica Chimica Acta*, vol. 403, 2009, pp.1-8

Bacon, Roger **Growth, Structure and Properties of Graphite Whiskers**, *Journal of Applied Physics*, vol. 31, 1960, p.283

Baptista, P. et al., **Gold nanoparticles for the development of clinical diagnosis methods**, *Anal Bioanal Chem*, vol. 391, 2008, p.943

Baringhaus, Jens et al., **Exceptional ballistic transport in epitaxial graphene nanoribbons**, *Nature*, vol. 506, 2014, p. 349

Baughman, Ray H., Anvar A. Zakhidov, Walt A. de Heer, **Carbon Nanotubes—the Route Toward Applications,** *Science,* vol. 297, 2002, p.787

Benyus, Janine M. **Biomimicry: Innovation Inspired By Nature,** Harper Collins Publisher Inc. New York, 1998, ISBN: 0-06-053322-6

Berry, S., H. Habarland, **Clusters of Atoms and Molecules vol. I** (Ed: H Haberland), *Springer Series in Chemical Physics,* vol. 52, 1994, p.1

Bethune, D. S. et al., **Cobalt-catalysed growth of carbon nanotubes with single-atomic-layer walls,** *Nature,* vol. 363, 1993, p.605

Binnig, G., H. Rohrer, Ch. Gerber and E. Weibel, **Surface Studies by Scanning Tunneling Microscopy,** *Physical Review Letters,* vol. 49, 1982, p.57

Binnig, G., H. Rohrer, Ch. Gerber and E. Weibel, **(7x7) Reconstruction on Si (111) Resolved in Real Space,** *Physical Review Letters,* vol. 50, 1983, p.120

Binnig, G., H. Rohrer, Ch. Gerber and E. Weibel, **Tunneling through a controllable vacuum gap,** *Applied Physics Letters,* vol. 40, 1982, p.178

Brown Univ. (RI) **Robotics: The future of minimally invasive heart surgery,** Department of Surgery, Course BI108, see website, 2012

Bryant, H. C. et al., **Magnetic needles and superparamagnetic cells,** *Phys. Med. Biol.,* vol. 52, 2007, p.4009

CBC News, **New Robot Technology Eases Kidney Transplants,** 2009 (July 8)

Chambliss, David D., Kevin E. Johnson (Goucher College), **Using STM to understand diffraction oscillations of Fe growth on Cu (100),** *Surface Science,* vol. 313, 1994, pp.215-26

Chen, Julian C. **Introduction to Scanning Tunneling Microscopy,** Oxford University Press, New York, 1993

Chen, J. et al., **Bright infrared emission from electrically induced excitons in carbon nanotubes,** *Science,* vol. 310, 2005, p.1171

Chihaia, V.,   S. Adams, W. F. Kuhs, **Influence of water molecules arrangement on structure and stability of 5^{12} and $5^{12}\,6^2$ buckyball water clusters. A theoretical study.** *Chemical Physics*, vol. 297, 2004, pp.271–287

Crommie, M. F.   C.P. Lutz, D.M. Eigler, E.J. Heller, **Waves on a metal surface and quantum corrals,** *Surface Review and Letters*, vol. 2 (1), 1995, pp. 127-137

Crommie, M. F.   C.P. Lutz, D.M. Eigler, **Confinement of electrons to quantum corrals on a metal surface,** *Science,* vol. 262, 1993, pp. 218-220.

Cristea, Ileana M.,   Simon J. Gaskell and Anthony D. Whetton, **Proteomics techniques and their application to hematology,** *BLOOD*, vol. 103, 2004, p.3624 (review)

Dean, Kenneth A.   **Industry Perspective, A new era: Nanotube displays***, Nature Photonics*, vol. 1, 2007, p.273

De Heer, W. A.   **The physics of simple metal clusters: experimental aspects and simple models,** *Review of Modern Physics*, vol. 65, 1993, pp.611-676 (review)

Demerjian, Dave   **Airplane Heal Thyself? Self-Repairing Aircraft Could Improve Air Safety,** *AUTOPIA* (online publication), 2008

Derfus, A. M.,   W. C. W. Chen, and S. N. Bhatia, **Probing the Cytotoxicity of Semiconductor Quantum Dots,** *Nano Letters*, vol. 4, 2004, p.11

Dong, W. G.   et al., **Research on properties of nano polypropylene/TiO_2 composite fiber,** *Journal of Textile Research*, vol. 23, 2002, p.22

Drexler, Eric K.   **Molecular Engineering,** *Proc. Natl. Acad. Sci. USA*, vol. 78 (no. 9), 1981 pp.5275-5278

Drexler, Eric K.   **Engines of Creation,** Anchor/ Doubleday, New York, 1986

Drexler, Eric K.   Nanorex; John Randall, Zyvex Labs; Stephanie Corchnoy, Synchrona; Alex Kawczak, Battelle Memorial Institute; Michael L. Steve, Battelle Memorial Institute (editors), **Productive Nanosystems: A Technology Roadmap,** *UT-Battelle, LLC under Contract No. DE-AC05-00OR22725,* U.S. Department of Energy, 2007 (open publication)

Dresselhaus, M.S., G. Dresselhaus and P. C. Eklund, **Science of Fullerenes and Carbon Nanotubes**, Academic Press USA, 1996

Endo, M. et al., **High resolution electron microscope observations of graphitized carbon fibers**, *Carbon*, vol. 14, 1976, p.133

Endo, M. et al., **Filamentous Growth of Carbon through Benzene Decomposition**, *J. Cryst. Growth*, vol. 32, 1976, p.335

Esfand, Roseita, Donald A. Tomalia, **Poly (amidoamine) (PAMAM) dendrimers: from biomimicry to drug delivery and biomedical applications**, *DDT*, vol. 6, 2001, p.427

Estey, E. P. **Robotic prostatectomy: The new standard of care or a marketing success?** *Canadian Urological Association Journal*, vol. 3, 2009, p.488

Fennimore, A. M., T. D. Yuzvinsky, Wel-Qlang Han, M. S. Fuhrer, J. Cumings and A. Zettl, **Rotational actuators bases on carbon nanotube**, *Nature*, vol. 424, 2003, p.408

Feynman, Richard **There's Plenty of Room at the Bottom—An Invitation to Enter a New Field of Physics**; Lecture, Christmas 1959. *Copyright*: California Institute of Technology, 1960. *Engineering and Science Magazine*, vol. XXIII (no.5), February, 1960.

Feynman, Richard **Infinitesimal Machinery—There is Plenty of Room at the Bottom, Revisited**—Lecture in 1983 as a sequel to 1959 lecture at Jet Propulsion Laboratories. *Journal of Micromechanical Systems*, vol.2, no.1, 1993, pp.4-14.

Finkelstein, J. et al., **Open versus Laproscopic versus Robot-Assisted Laproscopic Prostatectomy: The European and US Experience**, *Reviews in Urology*, vol. 12, 2010, pp.35-43

Fisher, A. J., P. E. Bloechl, **Adsorption and scanning tunneling microscope imaging of benzene on graphite and MoS2**, *Physical Review Letters*, vol. 70, 1993, p.3263

Fleger, J. E., L. Nifong, **The evolution of and early experience with robot assisted mitral valve surgery**, *Surg. Laprarosc Endosc. Percutan. Tech.*, vol. 12, 2002, pp.58-63

Fortina, P., L. Kricka, S. Surrey and P. Gordzinski, **Nanobiotechnology: the promise and reality of new approaches to molecular recognition**, *Trends in Biotechnology*, vol. 23, 2005, pp.168-73

Foroughi, Javad   et al., **Torsional Carbon Nanotube Artificial Muscles**, *Science*, vol. 334, 2011, p.494

Fréchet, J. M. J.   **Designing dendrimers for drug delivery**, *Pharmaceutical Science and Technology Today*, vol. 2, 2000, p.393

Fuhrmann, T.   et al., **Diatoms as living photonic crystals**, *Appl. Phys. B*, vol. 78, 2004, p.257

Fujishima, Akira,   Honda Kenichi, **Electrochemical Photolysis of Water at a Semiconductor Electrode**, *Nature*, vol. 238, 1972, p.37

Geim A. K.   and K. S. Novoselov, **The Rise of Graphene**, *Nature materials*, vol. 7, 2007, p. 183

Goldstein, G. I.,   D. E. Newbury, P. Echlin, D. C. Joy, C. Fiori, E. Lifshin, **Scanning Electron Microscopy and X-Ray Microanalysis**, Plenum Press, New York, ISBN 030640768X, 1981

Goldman, E. R.   et al., **Multiplexed toxin analysis using four colors of quantum dot fluororeagents**, *Anal Chem*, vol. 76, 2004, p.684

Gore, W. L.,   Gore-Tex, W. L. Gore and Associates GmbH, Germany

Grandbois, Michel,   Martin Beyer, Matthias Rief, Hauke Clusen-Schaumann and Hermann E. Gaub, **How Strong is a Covalent Bond?** *Science*, vol. 283, pp.1727-30

Graham P. Collins   **STM Rounds up Electron Waves at the QM corral**, *Physics Today*, vol. 46 No. 11, 1993, pp. 17-19

Gross, Leo,   Mohn Fabian, Moll Nikolaj, Liljeroth Peter and Mayer Gerhard, **The Chemical Structure of a Molecule Resolved by Atomic Force Microscopy**, *Science*, vol. 325, 2009, pp.1110-1114

Gutmman, Steven   **The Role of Food and Drug Administration Regulation of In Vitro Diagnostic Devices—Applications to Genetics Testing**, *Clinical Chemistry*, vol. 45, 1999, pp. 746-749

Hayter, Julia,   R. Duncan, H. L. Robertson, Simon J. Gaskell and Robert J. Beynon, **Proteome Analysis of Intact Proteins in Complex Mixtures**, *Molecular and Cellular Proteomics*, vol. 2.2, 2003, p.85

Heath, James R.,   **Nanoscience and Technology—A collection of Reviews from Nature Journals, Ed. Peter Rodgers**, World Scientific Publishing Co. under license, 2010, p. xi

Heath,James R.,   M. E. Davis, **Nanotechnology and Cancer**, *Ann. Rev. Med.*, vol. 59, 2008, pp.251-65

Hegemann, Dirk,   Michael Keller, Felix Reifler, Andri Vital, Giuseppino Fortunato, *Materials Science and Technology*, EMPA, Switzerland

Hirsch, I. R.   et al., **A whole blood immunoassay using gold nanoshells**, *Anal Chem*, vol. 75, 2003, p.2377

Huang, Xiaohua,   Wei Qian, Ivan H. El-Sayed, and Mostafa A. El-Sayed , **The Potential Use of Enhanced Nonlinear Properties of Gold Nanospheres in Photothermal Cancer Therapy**, *Lasers in Surgery and Medicine*, vol. 39, 2007, pp.747-753

Hyde, Kevin,   Mariana Rusa and Juan Hinestroza, **Layer-by-layer deposition of polyelectrolyte nanolayers on natural fibres: cotton**, *Nanotechnology*, vol. 16, 2005, p.S422-S428

IARC, Monographs **Evaluation of Carcinogenic Risks to Humans, Carbon black Titanium dioxide and Talc**, *World Health Organization (WHO)*, vol.93, Lyon, France, 2010

Ijima, Sumio   **Helical microtubules of graphitic carbon**, *Nature*, vol. 354, 1991, p.56

Ijima, Sumio,   Toshinari Ichihashi, **Single-shell carbon nanotubes of 1-nm diameter**, *Nature*, vol. 363, 1993, p.603

Inami, Nobuhito,   Mohd Ambri Mohamed, Eiji Shikoh and Akihiko Fujiwara, **Synthesis-condition dependence of carbon nanotube growth by alcohol catalytic chemical vapor deposition method**, *Science and Technology of Advanced Materials*, vol. 8, 2007, p.292

Ingall, Ellery   **Diatoms discovered to remove phosphorus from oceans**, Georgia Institute of Technology, Atlanta, USA, Published in Physorg website, 2008 (May 2)

IWGN, Report   **Shaping the World Atom by Atom: Document for non-professionals**, NSTC/CT, Govt. of USA, 1999

JLsun, High-Tech, Beijing JLsun High-Tech Co. Ltd., Caizhi International Mansion, No. 18, Zhongguancundonglu Road, Beijing City, China, P. C. 100083

Kanan, Matthew W., Daniel G. Nocera, **In Situ Formation of an Oxygen-Evolving Catalyst in Neutral Water Containing Phosphate and Co^{2+}**, *Science*, vol. 321, 2011, p.1072

Kataura, H.   et al., **Optical properties of single-wall carbon nanotube,** *Synthetic Metals,* vol. 103, 1999, p.2555

Kelvin, Lord   On the division of space with minimum partition area, *Phil. Mag. 5s,* vol. 24, 1887, pp.503-14

Kelly, Kevin   **The Next 5000 days of the web,** TED Talk, 2007 (December)

Kelly, Kevin   **What Technology Wants,** Viking Adult, 2010 (October 14)

Klostrance, J. M.   et al., **Convergence of Quantum Dot Barcodes with Microfluidics and Signal Processing for Multiplexed High-Throughput Infectious Disease Diagnostics,** *Nano Letters,* vol. 7, 2007, p.2812

Knight, Will   **Gecko tape will stick you ceiling,** *Article dn3785, New Scientist,* 2003 (June 1)

Knoll, Max   *Zeitschrift für Technische Physik,* vol. 16, 1935, p.467

Kumar, Mukul   Ando Yoshinori, **Carbon Nanotube from Camphor: An Environment friendly Nanotechnology,** *Journal of Physics: Conference Series,* vol. 61, 2007, p.643

Lanfranco, A. R.,   A. E. Castellanos, J. P. Desai, and W. C. Meyers, **Robotic Surgery: A current perspective,** *Annals of Surgery,* vol. 239(1), 2004, pp.14-21

Lin, F. Y.   et al., **Development of a nanoparticle labeled microfluidic immunoassay for detection of pathogenic micro-organisms,** *Clinical Diagnostic Lab Immunology,* vol. 12, 2005, p.418

Lipomi, Darren J.   et al., **Skin-like pressure and strain sensors based on transparent elastic films of carbon nanotubes,** *Nature Nanotechnology,* vol. 6, 2011, p.788

Losic, Dusan,   James G. Mitchell and Nicolas H. Voelcker, **Fabrication of gold nanostructures by templating from porous diatom frustules,** *New Journal of Chemistry,* vol. 30, 2006 p.908

Malik, N.   et al., **Dendrimer-platinate: a novel approach to cancer chemotherapy,** *Anti-Cancer Drugs,* vol. 10, 1999, p.767

Marks, Paul   **Self-powered displays keep gadgets alive,** *New Scientist, 2007, (May 5 issue), p. 30 Related*: US Patent No. 7206044 (Motorola), No. and 0080925 (Nokia)

Martin, T. P.   **Shells of Atoms,** *Physics Reports,* vol. 273, 1996, p.199

Marti, O., M. Amrein, (Ed.), **STM and AFM in Biology**, Academic Press, San Diego, California, 1993

Matthew, P. G., Lightcap Ian V. and Kamat Prashant V., **Sun-Believable Solar Paint. A Transformative One-Step Approach for Designing Nanocrystalline Solar Cells**, *ACS Nano*, vol. 6, 2012, pp.865-72

McConnell, P. I. E., W. Schneeberger and R. E. Michler, **History and development of robotic cardiac surgery**, *Problems in General Surgery*, vol. 20, 2003, pp.20-30

Medina, C. et al., **Nanoparticles: Pharmacological and toxicological significance**, *Br. J. Pharmacol.*, vol. 150, 2007, p.552

Melvin, W. S. et al., **"Robotic" Resection of A Pancreatic Neuroendocrine Tumor**, *Journal of Laproendoscopic and Advanced Surgical Techniques*, vol. 13, 2003, pp.33-36

Michalet, X. et al. **Quantum Dots for Live Cells, in Vivo Imaging, and Diagnostics**, *Science*, vol. 307, 2005, p.538 (review)

Mintmire, J. W., B. I. Dunlap and C. T. White, **Are Fullerene Tubules Metallic?** *Phys. Rev. Lett.*, vol. 68, 1992, p.631

Misewich, J. A. et al., **Electrically induced optical emission from carbon nanotube**, *FET, Science*, vol. 300, 2003, p.783

Mogilevsky, G., Q. Chen, A. Kleinhammes, Yue Wu, **The structure of multilayer titania nanotubes based on delaminated anatase**, *Chemical Physics Letters*, vol. 460, 2008, p.517

Morse, Daniel E. **Silicon biotechnology: harnessing biological silica production to construct new materials**, *Trends in Biotechnology (TIBTECH)*, vol. 17, 1999, p.230

Nam, J-M, C. S. Thaxton and C. A. Mirkin, **Nanoparticle based bio-bar codes for the ultrasensitive detection of proteins**, *Science*, vol. 301, 2003, pp.1884-86

NANO, Magazine **Nanotechnology and Textiles**, *NANO Magazine*, Issue No. 9, 2010 (January 5)

Naum Yvonne R., Edwin B. Matzke, **The Interrelationship between Orthic Tetrakaidecahedra and Rhombic Dodecahedra in Aggregation Series**, *Bulletin of the Torrey Botanical Club*, vol. 82, 1955, p.480

Nelson, David L.,   Michael   M.   Cox   (Ed.),   Lehninger   Principles   of
Biochemistry, 4th edition, ISBN: 1-4039-4876-3, 2005

Ohkoshi, Shin-ichi et al., Synthesis of a metal oxide with a room-temperature photoreversible phase transition, Nature Chemistry, vol. 2, 2010, p.539

Orbaek, Alvin W.,   Andrew C. Owens and Andrew R. Barron, Increasing the Efficiency of Single-Walled Carbon Nanotube Amplification by Fe-Co Catalysts Through the Optimization of C H4/H2 Partial Pressures, Nanoletters, vol. 11, 2011, p.2871

O'Regan, Brian,   Michael Grätzel, A low-cost, high-efficiency solar cell based on dye-sensitized colloidal TiO_2 films, Nature, vol. 353, 1991, p.737

Patel, Prachi   Water-Repelling Metals, technology review, published by MIT, Article ID# 21530, 2008 (October 15)

Popat, Ketul C. (Ed.) Nanotechnology in Tissue Engineering And Regenerative Medicine, Colorado State University, USA, CRC Press, ISBN. 9781439801413.

Qian, Lei   and Juan P. Hinestroza, Application of Nanotechnology for high performance textiles, Journal of Textile and Apparel, vol. 4, 2004) pp.1-7

Radushkevich,   Lukyanovich, Packing of C60 molecules and related fullerenes in crystals: A direct view, Soviet Journal of Physical Chemistry, 1952; Republished in. Chem. Phys. Lett., vol. 182, 1991, p.1

Raiesdana, S.,   M. H. Gholpayeghani and A. M. Nasrabadi, Biomimetic nanotechnology and nonlinear dynamics, IEEE Computer Society, 0-7695-2759-07/07, 2007

Read Leonard E.   Essay: I Pencil, My Family Tree as told to Leonard E. Read, Irvington-on-Hudson, New York: The Foundation for Economic Education

Reece, Steven Y.,   Jonathan A. Hamel, Kimberly Sung, Thomas D. Jarvi, Arthur J. Esswein, Joep J. H. Pijpers, Daniel G. Nocera, Wireless Solar Water Splitting Using Silicon-Based Semiconductors and Earth-Abundant Catalysts, Science, vol. 334, 2011, p.645

Reibold, M.,   P. Paufler, A. A. Levin, W. Kochmann, N. Patzke, D.C. Meyer, **Carbon Nanotube in an ancient Damascus sabre**, *Nature*, vol. 444, 2006, p.286

Ridley Matt   **Genome: The Autobiography of a Species in 23 Chapters**, 2010 (Harper Collins)

Roa, Wilson   et al. **Gold nanoparticle sensitize radiotherapy of prostate cancer cells by regulation of the cell cycle**, *Nanotechnology*, vol. 20, 2009, p.375101

Roco, M. C. (Chair) **IWGN Workshop Report**, *National Science and Technology Council*, Committee of Technology, Govt. of USA, National Science Foundation, 2004

Rohrer, H.   **The Nanometer Age: Challenge and Chance**, *Il Nuovo Cimento*, vol. 107, 1994, p.989

Rohrer, H.   **STM: 10 Years After**, *Ultramicroscopy*, vol. 42, 1992, pp.1-6

Russell, E.   **Nanotechnology and the shrinking world of textile**, *Textile Horizons*, vol. 9/10, 2002, p.7

Santoli, Salvatore   **Why nanostructuring and nonlinear informational dynamics for biomimicry?** *Nonlinear Analysis*, vol. 63, 2005, pp.e1299-e1309

Saxl, Ottilia   **Nanotechnology—What it means for the life sciences**, *Technology Nanobiotechnology*, Institute of Nanotechnology, UK

Seigel, R. W.   **Nanostructured Materials**, vol. 3, 1993, pp.1-18

Service, Robert F.   **Artificial Leaf Turns Sunlight into a Cheap Energy Source**, *Science*, vol. 32, 2011, p.25

Siebert, W.   et al., **Total knee replacement robotic assisted technique (Chapter 12), in Computer and robotic assisted hip and knee surgery** (ed. DiGioia, M. Anthony et al.), *Oxford University Press*, 2004 pp.127-156

Smith, B. R.   et al., **Localization to atherosclerotic plaque and biodistribution of biochemically derivatized superparamagnetic iron-oxide nanoparticles (SPIONs), contrast particles for magnetic resonance imaging (MRI)**, *Biomed. Microdevices*, vol. 9, 2007, p.719

Sokolov, K.   et al., **Real-time vital optical imaging of precancer using anti-epidermal growth factor receptor antibodies conjugated to gold nanoparticles**, *Cancer Research*, vol. 63, 2003, p.1999

Stroscio, J. A.,   W. J. Kaiser, (Ed.), **Scanning Tunneling Microscopy** Academic Press, Boston, 1993

Takesue, I., J.   N. Kobayashi Haruyama, S. Chiashi, S. Maruyama, T. Sugai and H. Shinohara, **Superconductivity in entirely end-bonded multiwall carbon nanotubes.** *Phys. Rev. Lett.*, vol. 96, 2006, p.57001

Talamini, M.,   W.C. Chapman, S. Horgan, W. S. Melvin, **Evaluation of 211 "Robotic" Surgical Procedures**, *Surgical Endoscopy*, vol. 17, 2003, pp.1521-24

Tanaka, R.   et al., **A novel enhancement assay for immunochromatograpic test strips using gold nanoparticles**, *Anal Bio Analytical Chemistry*, vol. 385, 2006, p.1414

Tennent, Howard G. **Carbon fibrils, method for producing same and composition containing same**, *Hyperion Catalysis International Inc.*, US Patent No. 4663230, 1987

Toffler, Alvin   **Future Shock**, BENTAM (3rd edition) 1984

Toumey, Chris   **Apostolic Succession**, *Engineering and Science*, No. 1/2, 2005

Turner, A. P. F.,   Beining Chen and Sergey A. Piletsky, **In-Vitro Diagnostics in Diabetes: Meeting the Challenge.** *Clinical Chemistry*, vol. 45, 1999, pp.1596-1601

Walsh, K. McNulty DOE, Brookhaven National Laboratory, USA

Wang, Duan L.   et al., **Rapid and simultaneous detection of human hepatitis B virus and hepatitis C virus antibodies based on a protein chip assay using nano-gold immunological application and silver staining method**, *BMC Infectious Diseases*, vol. 5, 2005, p.53

Wang, Q. H.   et al., **A nanotube-based field-emission flat panel display**, *Appl. Phys. Lett.*, vol. 72, 1998, p.2912

Watanuki, O.,   F. Sai, K. Sueoka, **Magnetic-force-sensing STM: novel application for simultaneous measurement of topography and field gradient of magnetic recording heads**, *Ultramicroscopy*, vol. 42, 1992, pp.315-20

Weiss, P. S.,   D. M. Eigler, Site dependence of the apparent shape of a molecule in scanning tunneling microscope images: Benzene on Pt {111}, *Physical Review Letters,* vol. 71, 1993, p.3139

Wiesendanger, R.,   H. J. Guntherodt (Ed.), Scanning Tunneling Microscopy, vols. I-III, *Springer Series in Surface Science,* vols. 20, 28 and 29, Springer-Verlag, Berlin, Heidelberg, 1992-93

Wong, Y. W. H.,   C. W. M. Yuen, M. Y. S. Leung, S. K. A. Ku and H. L. I. Lam, *AUTEX Research Journal,* vol. 8, 2006, p.191

Xiao, Yan   et al., Anti-HER2 IgY antibody-functionalized single-walled carbon nanotubes for detection and selective destruction of breast cancer cells, *BMC Cancer,* vol. 9, 2009, p.351

Xing, Yun,   Rao Jianghong, Quantum dot bioconjugates for in-vitro diagnostics and in vivo imaging, *Cancer Biomarkers,* vol. 4, 2008, pp.307-319

Yasuhide, Y.   et al., Composite material carrying zinc oxide fine particles adhered thereto and method for preparing same, *EP Patent No. 0791681,* 1997

Yuan, Shoucai,   Yuan Shouhuai, Hong Yingfang and Zhu Changchun, Carbon nanotubes flat panel displayer fabricated and its high voltage drive circuits designed, *IEEE, Photonics and Optoelectronics Symp.,* Chengdu, China, 2010 (19-20 June)

Zhang, J.   et al., Hydrophobic cotton fabric coated by a thin nanoparticulate plasma film, *Journal of Applied Polymer Science,* vol. 88, 2003, p.1473

Zhou, W. Z.   et al., Studies on the antistatic mechanism of tetrapod-shaped zinc oxide whisker, *Journal of Electrostatics,* vol. 57, 2003, p.347

Websites

http://nobelprize.org/nobel_prizes/physics/laureates/1986/press.html (Nobel com/autopia/2008/05

http://www.technologyreview.in

http://www.physorg.com/print128959346

http://www.cbid.gatech.edu/biomimetic.html

http://en.wikipedia.org/wiki/Robotic_surgery

http://biomed.brown.edu/Courses/BI108/BI108_2000_Groups/Heart_Surgery/Robotics.html (site: Brown University, RI, USA)

http://esciencenews.com/articles/2008/11/20/

http://wcbstv.com/health/da.vinci.robot.2.1055154.html

http://en.wikipedia.org/wiki/Da_Vinci_Surgical_System

http://www.nature.com/news/2006/061113/full/news061113-11.html (Nature Journal)

http://www.biomedcentral.com/1471-2407/9/351 (BMC Cancer open access)

http://www.mansolar.com (DSSC Solar Kit)

www.acsnano.org (Open access: Sun-Believable Solar Paint)

http://www.tatasteeleurope.com/en/news/2011_dscc

http://www.eurekalert.org/pub_releases/2007-05/ns-sdt050207.php (NOKIA self-powered display)

http://www.nanotex.com

www.autexrj.org (AUTEX Research Journal)

http://www.empa.ch (EMPA, Switzerland)

http://www.jl-chem.com/content/eAnti-bacterial.htm (Beijing JLson Hi-Tech Co.)

http://www.nano.org.uk/articles/28/ (NanoMagazine)

http://www.nanowerk.com/spotlight/ (Nanotechnology textiles)

http://www.hohenstein.de (Hohenstein Quality Control in Nanotextile, Germany)

http://www.toray.com (NanoMATRIX, TORAY Industries Ltd.)

http://www.sensatex.com (SENSATEX, Bethesda, MD USA)